Herbert William Page

Injuries of the spine and spinal cord without apparent mechanical lesion, and nervous shock

in their surgical and medico-legal aspects. Second Edition

Herbert William Page

Injuries of the spine and spinal cord without apparent mechanical lesion, and nervous shock
in their surgical and medico-legal aspects. Second Edition

ISBN/EAN: 9783744738439

Printed in Europe, USA, Canada, Australia, Japan

Cover: Foto ©berggeist007 / pixelio.de

More available books at **www.hansebooks.com**

INJURIES

OF THE

SPINE AND SPINAL CORD

WITHOUT APPARENT MECHANICAL LESION,

AND

NERVOUS SHOCK,

IN THEIR

SURGICAL AND MEDICO-LEGAL ASPECTS

BY

HERBERT W. PAGE, M.A., M.C. Cantab.,
FELLOW OF THE ROYAL COLLEGE OF SURGEONS OF ENGLAND;
SURGEON TO, AND LECTURER ON SURGERY AT, ST. MARY'S HOSPITAL FORMERLY
SURGEON TO THE CUMBERLAND INFIRMARY

SECOND EDITION

LONDON
J. & A. CHURCHILL
NEW BURLINGTON STREET
1885

PREFACE TO THE SECOND EDITION

THE favourable reception accorded to the first edition of this book both in our own country and abroad has called for a second edition so much sooner than I had ventured to hope, that although in preparing this edition I have subjected the work to thorough revision, I have been unable to make such additions to it as I had wished and contemplated. I have, however, availed myself of some valuable suggestions in the writings of those who, more especially in America, have been led by a perusal of the first edition to contribute to periodical literature their own experience upon the subjects dealt with in this book, and whose views I am glad to find very largely coincide with those which I have submitted to the profession.

In its main features the work remains unchanged, and my hope is that it may still be found to contain a trustworthy record of the various injuries and diseases which are caused most commonly, but by no means exclusively, by railway collision.

146, HARLEY STREET,
January, 1885.

PREFACE TO THE FIRST EDITION

IT has fallen to my lot in the past nine years to have seen, in the capacity of Surgeon to the London and North Western Railway Company, a large number of injuries about which but little information is to be learned in the text-books of medicine and surgery. I desire, therefore, to bring before the profession the results of my observation and experience, in the hope that I may to some extent succeed in throwing light upon much that is obscure, and in helping others to a clearer, and I would fain trust a more correct, view than has hitherto been afforded of the injuries, and the consequences of the injuries, received in railway collisions.

If to any it should seem impossible that a work upon railway injuries, surrounded as they too often are by elements happily absent from the more ordinary cases of medical practice, can be written in any spirit other than of partiality and bias, I would distinctly state that although my experience has been for the most part gained while acting as surgeon to a railway company, there has been nothing whatever in the circumstances of my appointment to impair, even in the very smallest degree, that free and perfect independence which is the rightful possession of a medical man. More than this it is unnecessary to say.

Parts of this book, that is to say, Chapter I on Concussion of the Spinal Cord; Chapter II on "Concussion of the Spine"; Chapter III on the Common Spinal Injuries of Railway Collisions, and much also of Chapter VIII, formed the dissertation upon "Injuries of the Back, without apparent Mechanical Lesion, in their Surgical and Medico-Legal Aspects," to which in 1881 was awarded the Boylston Medical Prize of Harvard University, U.S. By an order of the Boylston Committee of 1826 it is requisite to print here the following votes:

> 1st. "That the Board do not consider themselves as approving the doctrines contained in any of the dissertations to which premiums may be adjudged."
>
> 2ndly. "That in case of publication of a successful dissertation the author be considered as bound to print the above vote in connection therewith."

I am greatly indebted to my friends Dr H. B. Donkin and Dr John Macdougall, of Carlisle, for many valuable suggestions, and for much help given me in the preparation of this work; but not less sincerely must I thank those, too numerous to be named, who have by kindly interest and inquiry enabled me to compile the Appendix of Cases, and without whose help that Appendix must have been imperfect and almost valueless.

146, HARLEY STREET.
Dec., 1882.

CONTENTS

CHAPTER I

CONCUSSION OF THE SPINAL CORD

                                                                                                            PAGE

I. Gravity of injuries to the spinal cord—Supposed effect of railway collisions upon the cord—Brodie's classification of injuries to the cord—Intraspinal hæmorrhage—Cases of concussion lesion by Brodie, Abercrombie, Bell, Mayo, Boyer, Syme, Lidell, Edmunds—Danger of spinal cord lesions . . . . . . 1—23

II. Analogy between concussion of brain and concussion of spinal cord—Lesions in brain concussion—Views thereon and cases of Hewett and Hutchinson—Classification of concussion injuries of brain—Physical surroundings of brain and spinal cord—Views of Savory—Further cases of concussion of spinal cord by Abercrombie, Lidell, Hutchinson, Shaw, Gull, Mayo, Curling, Lochner, Hunter, Sharkey, Bastian—Conclusions . . . . 23—57

CHAPTER II

"CONCUSSION OF THE SPINE"

I. Mr Erichsen's works on railway injuries—Objections to the term "concussion of the spine"—Mr Erichsen's views on railway injuries—His typical case of "concussion of the spine"—Bell's observations on the same—Danger of traumatic meningitis—Syphilitic meningitis—Cases recorded by Mr Erichsen—Remarks thereon . . 58—82

II. Mr Erichsen's teaching on "concussion of the spine" and jar of the cord from slight or indirect injury—Analogy of the broken watch and fractured bones—Symptoms of

b

"concussion of the spine" from indirect injury—Pathology of "concussion of the spine"—Mors silet—The hysterical paralytic—Mr Erichsen upon the absence of certain *post-mortem* knowledge of railway injuries—Examination of his reasons—Chronic meningitis and subacute myelitis—Mr Gore's case of railway injury—Lockhart Clarke's examination thereof—Mr Shaw upon the same—Views on "concussion of the spine" and spinal cord by Shaw, Le Gros Clark, Erb .   82—112

CHAPTER III

THE COMMON SPINAL INJURIES OF RAILWAY COLLISIONS

Frequency of injuries of the back in collisions—Mode of occurrence—Cases—Nature of the injury—Association with nervous shock—Spinal pain—Pseudo-paralysis—Spinal tenderness—Difficulty in micturition and constipation—Peripheral sensations after spinal injury, and Hilton's views thereon—Injury to central nervous structures—Pachymeningitis, and syphilis as a cause thereof—Spinal caries—Symptomatic value of spinal pain—Shaw, Wilks and Gowers upon the same—Treatment of spinal pain and stiffness—Hyperæsthesia of the back—Prognosis of spinal injuries—Traumatic origin of tabes dorsalis, and Dr Petit upon the—Influence of compensation   113—157

CHAPTER IV

SHOCK TO THE NERVOUS SYSTEM

Nature of shock or collapse—Fright a potent cause thereof—Especially in railway accidents—Different degrees of collapse—Deferred shock—Frequent absence of bodily injury—Cases—Shock at different ages—Symptoms of general nervous shock—Sleeplessness—Disturbances of circulation—Headache—Nervousness—Sweating—Vasomotor paresis—Asthenopia—Loss of memory—Ovarian derangements—Effects of railway concussion on pregnancy—Death from uncomplicated shock—Mr Hutchinson on the symptoms of brain concussion .   158—187

CHAPTER V

SHOCK TO THE NERVOUS SYSTEM (*continued*)

PAGE

Condition of patients—" Hysteria " in men—Tendency towards gradual convalescence—Delayed recovery—Combination of mental and physical causes inducing delay—Alarm—The hysterical state—The organic sensations, Sully and Maudsley upon—Physical basis of abnormal sensations, Paget upon—Concentrated attention and anxious reflection—Baneful influence of litigation—Causes of " chronic invalidism "—Continued nervous shock absent in cases of severe fracture—Explanation thereof—Misuse of bromide of potassium—Chances of recovery—The nervous temperament—Imperfect recovery—Maintenance of symptoms—Spinal anæmia—Ophthalmic changes in spinal injuries, Gowers and Bruce-Clarke on   188—211

CHAPTER VI

FUNCTIONAL OR NEUROMIMETIC DISORDERS

The mimicry of nervous diseases—Predisposition thereto—Higher intellectual processes *hors de combat*—Wilks on arrest of cerebral action—Diagnostic errors in consequence of mimicry—Nature of functional changes—Hyperexcitability and paresis—The hypnotic state more or less under control—Hypnotic phenomena injurious to the nervous system—Cases—Moxon on the nutrition of the spinal cord—Functional paraplegia—Abeyance of the will—Cases in men—Symptoms not dependent on chronic meningitis or subacute myelitis   .   212—245

CHAPTER VII

MALINGERING

The simulations of disease—Ogston's classification—Hilton on deception—Tympanitis—Hæmorrhage—Use of atropine—Feignings of paralysis—Precedent diseases used for purposes of fraud—Asymmetry—Emotional disorders not necessarily feigned—Examples—Malingering after railway accidents—The motive—Its prevalence and influence—Induced malaise—" Spurious nervous shock " and "concussion of the spine"—Vomiting—Assumption of spinal injury—Functional disorders—Hysterical hemianæsthesia—The bearing of individuals . 246—278

CHAPTER VIII

CONCLUDING REMARKS

Influence of compensation on clinical features of disease—Illustrative case—Leading questions—Leading examinations—Uses of electricity—The " electric test "—Methods of clinical investigation—Cross-examination—Medical evidence in courts of law—The medical witness 279—295

APPENDIX OF CASES . . . 296

INDEX OF SUBJECTS . . 379

INDEX OF AUTHORS 399

INJURIES OF THE SPINE AND SPINAL CORD

CHAPTER I

CONCUSSION OF THE SPINAL CORD

I

The structural integrity of the central nervous system is of supreme importance in the animal economy; and it may with truth be said that of all the accidents to which man is liable, none more serious can befall him than injury to the spinal cord.

There is, indeed, no organ in the human body whose structural damage is fraught with such grave results; for whether we regard it as the conductor of impressions to and from the brain, as the co-ordinating centre of orderly automatic movements, or as the controlling guardian of healthy nutrition and change, the integrity of this part of the cerebro-spinal axis is essential for the due performance of its varied functions. Were the spinal cord a shapeless mass like the liver or the spleen, it is conceivable that severe damage might be done to it without interference with its peculiar functions; but, being as it is a slender cord, injury, like the broken link in a chain, renders its function inert, revealing how the integrity of the whole depends upon the integrity of a part. To have the use of the limbs impaired, or altogether lost ; to have them cut off, as it were, from the sentient life of the organism, is an infliction which, whether the result of injury or disease, very rightly awakens the sympathy of mankind.

It will not be denied, we think, that there is a very widespread impression that the spine and spinal cord are liable, before all other parts, to meet with injury in railway collisions; and under the name " concussion of the spine " is classed a number of complaints and symptoms which are commonly met with after railway accidents, but which are rarely seen or heard of in the ordinary practice of the profession. Many cases had not come under our observation before we found that "concussion of the spine" was assumed almost as a matter of course to have befallen the sufferer in a railway collision, and our attention has thus been especially directed to the important subject of injuries to the spine and spinal cord.

It is right, therefore, that we should seek, in the very outset of our inquiry, to learn how far the spinal cord is really liable to injury; for if the current views upon this subject be erroneous, it is of utmost importance that our patients should be saved from an overwhelming and very natural anxiety, which in that case would have no sure foundation of clinical and pathological data whereon to stand.

The difficulty of the inquiry would be materially lessened had we to treat of the grosser lesions of the spinal column and its contents with which all are familiar, and of which the wards of every hospital from time to time contain examples. It is not, however, of fracture of the spine or dislocation of the vertebræ that we here must speak. These are very commonly accompanied by serious lesion of the spinal cord; and the symptoms, and too often the early death, remove all doubt as to the nature of the injury sustained.

The lesions which collision accidents are presumed to cause are far more subtle. The accident has been slight; there is no evidence of any blow having been received; the injured person is unaware, perhaps, that he has been hurt, and yet, nevertheless, the delicate structure of the spinal cord is supposed to have been damaged by the jar of the collision, and to have become liable thereby to morbid structural changes, which, not manifest at first, most surely advance as time goes on, revealing themselves by symptoms only after months or years, and dooming the sufferer to a life of pain, misery and uselessness.

This is the hopeless prospect which may be so often heard foretold, and it compels us to make full inquiry into the liability of the spinal cord to suffer accidental lesion, that we may clearly comprehend what is the real nature of the common injuries received in railway collision which become so often the subject of medico-legal investigation.

We shall do wisely, in the first place, to examine for ourselves into the injuries to which the spinal cord is liable in the ordinary accidents of every-day life, and more especially into those cases which, from an assumed analogy with concussion of the brain, have been recorded as examples of concussion of the spinal cord.

If we turn to Sir Benjamin Brodie's paper on "Injuries to the Spinal Cord"* we there learn from him that the following is a "not inaccurate representation" of the injuries which are met with, and which his readers must themselves have witnessed, in civil and military practice.

" I. Fractures without displacement.

" II. Fractures with depression or displacement, and causing pressure on the spinal cord.

" III. Fractures complicated with dislocation.

" IV. Dislocations not complicated with fracture.

" V. Extravasations of blood on the surface of the mem-

* 'Med.-Chir. Trans.,' vol. xx, p. 120. Here abbreviated.

branes of the spinal cord," which "rarely take place to any considerable extent, and bear no comparison to those which occur within the cavity of the cranium."

" VI. A narrow clot of extravasated blood is sometimes discovered within the substance of the spinal cord. . . . .

"VII. Laceration of the spinal cord and its membranes.

" VIII. The minute organisation of the spinal cord may suffer from a blow inflicted upon the spine, even where there is neither fracture nor dislocation, and where the investing membranes do not appear to participate in any way in the effects of the injury." "In such cases," he goes on to say, " if there be an opportunity of examining the spinal cord at a very early period after the accident has occurred, the central part of it is found to be softer than natural, its fibrous appearance being lost in that of a semi-fluid substance. If the patient survives for a longer period, the alteration of structure is perceptible in the whole diameter of the cord, and occupies from one to two inches, or even more, of its length; and at a still later period it has often proceeded so far as to terminate in its complete dissolution.

"This disorganization, softening, and final dissolution of the spinal cord, is the most common consequence of injuries of the spine, and the dangerous symptoms which follow these accidents are, in the majority of cases, to be attributed to it. It bears no distant resemblance to the effects of a contusion of those soft parts which are more superficially situated, and it is easy to understand that it may be produced by a severe concussion operating on the delicate medullary fibres and cineritious substance, of which the spinal cord is composed."

We must especially direct attention to this the last of Brodie's divisions, for it bears upon the following passage where he says that " the effect of a violent concussion is at once to impair and even to destroy the function of the spinal cord, sometimes even causing the patient's death in the course of a few hours; and the question here presents itself, what is the nature of the injury thus inflicted on the spinal cord, so trifling in appearance, so great in reality, which is capable of producing such important and dangerous consequences?"* In answer to this question he combats the opinion of those who hold that this softening and dissolution of the cord are the consequence of inflammation, by pointing out that, firstly, it may be detected before there has been time for inflammation to arise; secondly, because there is no appearance of increased vascularity; thirdly, because even in advanced cases where there is complete disorganisation the membranes are for the most part natural, free from vascularity, lymph, serum, or pus; and fourthly, because the "symptoms which mark the progress of these changes are merely a continuation of those which the concussion of the spinal cord has occasioned in the first instance, and which of course must have been wholly unconnected with inflammation." It is remarkable, however, that Brodie gives no case where the cord has been proved post-mortem to have undergone the changes described as the sole result of a blow inflicted upon the spine; and if we examine the cases which he does bring forward, their evidence appears to us inadequate to establish the doctrine which he here lays down. Thus, in recording the history of a patient "whose lower limbs were paralytic after a severe blow on the spine," and who "regained the use of them in the course of three or four weeks," he remarks that "it is easy to understand that where paralysis is produced by the pressure of extravasated blood, it may be relieved by the absorption of the coagulum; or that the injury inflicted by concussion on the structure of the spinal cord, may be gradually repaired."† We cannot doubt that the former of these two explanations is the one to be received. And in recording another case, where *immediate* paraplegia from a blow on the back, in a

* P. 124. † P 131.

man aged forty-five, was followed in nine days by "severe cramps and painful convulsions," and where, at the necropsy, "the membranes of the spinal cord, and the spinal cord itself, presented a natural appearance externally; but on the latter being divided longitudinally, the central part of it was found to be in a softened state, so that on being macerated for a short time in water, it almost completely disappeared," Brodie does not appear—if we may venture to question the teaching of so illustrious a surgeon—to lay sufficient stress upon the fact that there had been "a fracture of the fourth dorsal vertebra with such a degree of displacement as to produce a slight degree of pressure on the spinal cord."* That he does not lay due stress upon it is obvious, we think, from the fact that in commenting on this case, and on another case, not fatal, where there were symptoms of muscular spasm with paraplegia, and where there were "fracture and considerable displacement of the third and fourth lumbar vertebræ," he writes: "I am the more inclined to believe that this (*i.e.* some degree of pressure on the spinal cord) was the cause of the spasmodic affection of the muscles, as I have not met with any case in which it was proved by dissection that this symptom existed in combination with disorganisation of the cord, and, independently of pressure on it."† Brodie would thus appear to look upon muscular spasm as an undoubted symptom of pressure on the spinal cord, and to regard the disorganisation of the cord as a result of direct concussion blow, by the same blow in fact which had perhaps caused fracture and displacement; but he gives no example of disorganisation without the existence also of fracture or displacement. Two of these three cases recovered: the one perfectly, as far as we know, in whom we should have little doubt that the paralysis was produced by the "pressure of extravasated blood" which was ultimately absorbed; the other was lost sight of after he left the hos-

* P. 134. † P. 135.

pital much improved: but in that case also we should say that there was hæmorrhage, ultimately absorbed, with probable damage by the displacement to the membranes and the nervous cords; for it must be borne in mind that "fracture and considerable displacement of the third and fourth lumbar vertebræ" would cause damage not to the spinal cord but to the nerves of the cauda equina. Had the cord itself been disorganised in either of these cases the patients would not, in our judgment, have recovered. Throughout the whole paper the possibility seems to be assumed of "disorganisation and dissolution of the cord" being the result of "concussion," and that "the injury inflicted by concussion on the structure of the spinal cord may be gradually repaired," but it is worthy of prominent remark that no case is recorded which unequivocally supports the opinions expressed in the paper.

Quoting some remarks on the nature of the injuries in concussion of the brain, from a previous contribution of his own ('Med.-Chir. Trans.,' vol. xiv), he says that we are not justified in the conclusion that because no changes are to be detected after death in cases commonly called concussion of the brain, "there is therefore in reality no organic injury." "There may be changes and alterations in it which our senses are incapable of detecting. The speedy subsiding of the symptoms in some cases of concussion does not contradict this opinion. A deep incised wound in other parts may, under certain circumstances, be completely and firmly united in the space of twenty-four hours, and it is easy to suppose that the effects of a much slighter injury may be repaired in a much shorter space of time;" and then he goes on: "These remarks are not less applicable to cases of concussion of the spinal cord than they are to those of concussion of the brain. We cannot doubt that the nature of the injury is the same in both of them. It is true that much worse consequences usually arise from concussion of the spinal cord than from concussion of the brain; that is, if the patient recovers, his recovery is more tedious; that if he dies, greater changes in the condition of the injured part are detected on dissection."* The difference is easily to be explained by the different mechanical relations of each organ to the bony walls around it. Even, however, if the analogy hold good—and this we shall discuss later—between concussion of the brain and concussion of the spinal cord, it is, to say the least, very doubtful whether any passing paralysis following a severe blow on the vertebral column, is not much more likely to be due to the pressure of extravasated blood, which in course of time becomes absorbed. And it is very noticeable that he does not record one single case of supposed concussion of the spinal cord *which ended fatally*, and where there was not also serious injury to the spinal column. Bearing upon this point we must not omit Brodie's remark that "in reviewing the various consequences of injuries to the spinal cord, we find nothing more remarkable than the following circumstance: that whether the cord be lacerated or compressed, or has undergone that kind of disorganisation which is induced by a severe concussion, there is no material difference in the symptoms which arise, or in the results to which they lead."† A noteworthy statement, because it is clear that he had in his mind cases where injury was immediately followed by the most serious symptoms, and where there was little or no tendency towards recovery. Had the patient, whose case we have previously quoted, by some mysterious chance recovered, or had there been no necropsy when he died, we should never have known of the "fracture of the fourth dorsal vertebra with such a degree of displacement as to produce a slight degree of pressure on the spinal cord;" and we need, it seems to us, better evidence than Brodie adduces to establish the existence of such a pathological state as

* Vol. xx, p. 123.   † P. 154.

"disorganisation and dissolution of the cord" being produced by concussion pure and simple, without some serious mechanical injury being inflicted upon the spinal column as well. We now know, moreover, that in fracture of the spine, or in separation of vertebræ without fracture, the spinal cord may, at the very moment of injury, have been severely crushed by displacement of the bones, a displacement no longer discernible during life because the vertebræ have started back immediately into their natural positions. Examples of this most surgeons must have seen.

In his well-known work on 'Diseases of the Brain and the Spinal Cord'* a section is devoted by Abercrombie to "Concussion of the Spinal Cord," wherein he writes, "A severe blow upon the spine frequently occasions an immediate loss of power in the parts below the seat of the injury without producing either fracture or dislocation of the vertebræ." It is most remarkable, however, that Abercrombie, whose work is essentially practical, and contains the record of many interesting cases, gives only one instance of such a condition seen by himself; and that this opinion is apparently founded upon published cases of Boyer and Frank and others. To his own case we shall presently refer. Further, in speaking of concussion of the spinal cord, he says, "It may produce permanent paralysis. This may occur immediately, or the first effects of the injury may be recovered from, and a new diseased action may take place after a considerable time. . . . . The slight nature of the first symptoms, in such cases, and the slowness of their progress, will be illustrated by the following case," which we shall here abbreviate : "Case CLI, a man aged fifty-four, about twenty-five years ago fell from the branch of a tree and lighted on the sacrum. He was carried home complaining of pain in the lower part of the spine, and entirely paralytic in his lower extremities. In this state he was confined to bed about twelve days, and

* Fourth edition, section ix, p. 372, *et seq.*

then recovered so as to be able to follow his usual employment; but from this time he was affected with a peculiar feeling of numbness, which was confined to the upper part of the left foot." He continued in this state for four years, and then the numbness extended upwards, and he had pain across the lower part of the back, and some palsy of the right thigh and leg. He was again in bed for two years with paraplegia. He then recovered a little, but ten years after the accident there was still partial paraplegia and incontinence of urine. Such is the outline of the case, but we think that any one who carefully examined the history would be perfectly justified in entertaining very considerable doubt whether hæmorrhage associated with undiscovered fracture, or even without fracture, was not a much more probable cause of the symptoms—both immediately, and secondarily as affecting the membranes or nerves—than any direct damage to the structure of the spinal cord by concussion blow. Nor could the early symptoms be regarded in any sense of the word as "slight," when on being carried home the man was found "entirely paralytic in the lower extremities."

Sir Charles Bell, in his 'Institutes of Surgery'[*] records cases which, both in the mode of injury and in the after symptoms, bear a very striking resemblance to this last case of Abercrombie; and he remarks that the symptoms are probably to be explained by some "injury to the soft envelope of the spinal marrow, and the accession of inflammatory thickening."

In his 'Outlines of Pathology'[†] Mayo makes reference to the same subject, and says that concussion of the spinal marrow may produce complete suspension of its functions. We quote:[‡] "The following cases exemplify the effects of simple concussion of the spinal marrow; in the second,

---

[*] 'Institutes of Surgery,' 1838, vol. i, p. 154, *et seq.*
[†] H. Mayo's 'Outlines of Human Pathology,' 8vo, 1836.
[‡] P. 156, *et seq.* Italics not in original.

strong threatenings of inflammatory softening of the cord manifested themselves. A man, aged fifty, was admitted between four and five years ago into the Middlesex Hospital, having fallen out of a loft into a stable, in such a manner as to *pitch upon the juncture of his head and neck.* He did not lose his senses, but, on being lifted up, his arms and legs were found to be numb and powerless. In a few days he recovered the feeling and use of his legs; the numbness and weakness likewise gradually left the arms, but his hand remained affected and continues so still. .

. . J. J., aged thirty-nine, admitted in May, 1835. On the first of March he was turning to speak to some one on the top of a flight of seventeen stone steps when he slipped and fell backwards to the bottom. He was stunned by the fall, but knows that he pitched upon the upper part of the back, because his coat was cut through at this part, and his shoulders and back were bruised. He was lifted up and soon recovered and walked to his room. No symptoms supervened for a month, during which he recovered from the bruise, and lived heartily as before the accident. He was then, without any warning, seized with spasm in the left foot and hand; the spasm went off in a few minutes, but the left hand and arm remained weak and numb. This was attended with pain of the back part of the head, and occasional confusion of thought, and aching and shooting pains between the shoulders.; he had also frequent desire to make water, which came on suddenly with great urgency. He continued in this state about a fortnight, when he had twitchings in the arm and leg, and gradually recovered the use first of the arm and then of the leg. A fortnight after the restoration of his arm and leg his right side was taken much as the left had been. This again got better. At the time of, and for six weeks after, his admission, he was liable to spasmodic seizures of the hands and feet, which lasted a few minutes; the pulse during the seizure was frequent and feeble; the skin cold and inclined to rigor; his limbs were weak, and he had pain at the back of the head, and occasional confusion of thought. On striking the upper dorsal vertebræ, an obscure and deep-seated pain was felt in the part. He has now (August 17th) for several weeks been entirely free from symptoms. He has remained in the hospital; been cupped upon the back, had issues applied over the part which was struck; and for six weeks the mouth was kept slightly affected with mercury." Neither of these cases appears to us at all conclusive. Looking to the peculiar nature of the injury, and the whole history of the symptoms in the first case recorded, our impression would certainly be that *concussion* of the spinal marrow *per se* was just the very condition which did not give rise to them, but that they were due rather to slight hæmorrhage which pressed upon the spinal marrow, or to some injury to the cord itself by the severe bend which in all probability it received when the patient pitched upon the juncture of his head and neck. Of the second case it may be remarked that it is altogether exceptional; nor is there anything indeed to show that the symptoms were due at all to injury to the spinal cord. Assuming that they were, we know nothing as to the possible existence of syphilis, a fertile source of error in all such cases. But note the history. There were no symptoms for a month, and then the man without any warning was suddenly seized. Were there no cerebral symptoms? Might not all be better explained by cerebral embolism? Can, forsooth, the case be accepted as an indubitable example of concussion lesion of the spinal marrow?

Boyer* writes as follows upon the same subject: "Toute percussion violente portée sur l'épine, qu'elle produise ou non la fracture de quelqu'une des parties des vertèbres, ne borne pas ses effets à la colonne vertébrale. L'ébranlement se communique à la moelle de l'épine, et peut produire sur cet organe délicat les mêmes effets que sur le cerveau.

* 'Maladies Chirurgicales,' cinquième édition, tom. 3, p. 133, *et seq.*

Ces effets sont beaucoup plus considérables et plus à craindre quand la fracture intéresse la lame d'une ou de plusieurs vertèbres, et que les fragments dirigés vers l'intérieur du canal vertébral ont lésé la moelle épinière ou ses enveloppes, ou qu'ils compriment ces mêmes parties. Ces complications qui accompagnent fréquemment les lésions de la colonne vertébrale méritent toute l'attention du praticien, et sont beaucoup plus graves que la fracture elle-même. On voit alors survenir, ou sur-le-champ, ou quelque temps après l'accident, selon qu'il a produit une fracture avec enforcement, une commotion ou un épanchement sanguin; on voit, dis-je, survenir, une paralysie complète ou incomplète des extrémités inférieurs, de la vessie et du rectum." He records the following cases:* "Un ouvrier en bâtiments tomba d'environ quatorze pieds d'élévation et perdit connaissance. Revenu a lui, il s'aperçut qu'il avait perdu l'usage des extrémités inférieurs; l'urine était retenue dans la vessie, les matières fécales l'étaient pareillement d'abord, et puis s'échappèrent involontairement. La fièvre survient, la respiration devient laborieuse, et le malade succombe le douzième jour de l'accident. A l'ouverture du cadavre, nous trouvâmes un épanchement de sérosité sanguinolente qui remplissait le canal de la dure-mère, depuis sa partie inférieure jusqu'au milieu du dos, et qui comprimait la moelle épinière." . . . "Un ouvrier fabricant de bas tombe sur les reins dans un fossé profond, et se trouve aussitôt paralysé des extrémités inférieurs, de la vessie et du rectum. La maladie suit la même marche que dans les cas précédents, et le malade ne tarde pas à succomber. A l'ouverture du cadavre, nous ne trouvons ni fracture, ni lésion de la moelle épinière ou de ses envelloppes, ni épanchement." . . . "Un homme s'amusant avec ses amis à faire des tours de force dans une posture difficile, éprouva un tiraillement violent et une douleur aiguë dans la longueur de l'épine. Le

* Op. cit., p. 135.

lendemain, les membres inférieurs, la vessie et le rectum furent paralysés; la maladie suivit la marche accoutumée, et le malade mourut au bout de quelques semaines. L'examen de son cadavre fit voir les parties dans leur état naturel comme dans le cas précédent."

It undoubtedly appears strange that in these cases, with the exception of the first, where there was ample evidence of the real cause of the paralysis—and let us note by the way that that was hæmorrhage outside the cord—no structural changes were found after death. We must remember, however, that the methods of post-mortem examination were then far different from what they are now. With such meagre histories as are here recorded we shall not, it seems to us, be hypercritical in refusing to accept these often-quoted cases as evidence that the spinal marrow may suffer lesion from the concussion pure and simple of a blow. On whatever cause the palsy may in the two last cases have depended, suffice it to remark that there is no evidence of this having been structural lesion of the spinal cord. True, it may be said that these were cases in which the function of the spinal cord was annihilated by the blow, as the function of the brain may be annihilated by a severe blow upon the head; but if that be so, it is a wholly remarkable circumstance that in serious injury to the spinal column, where there is crush of the spinal cord, death does not invariably ensue long before that extension of the paralysis which experience tells us is the usual ending of cases of fracture and dislocation of the spine.

As more strictly in accordance with clinical and pathological experience we may quote the following passage from Syme:* "*Injuries and diseases of the spinal canal and its contents.* The spinal cord is liable to concussion from blows and falls, particularly the latter, the symptoms of which are similar to those of concussion of the brain, inasmuch as they denote suspension of the functions usually

* 'Principles of Surgery,' 3rd edition, p. 433.

exercised by this part of the nervous system. As these consist chiefly in conduction of the impressions producing sensation and voluntary motion, the patient loses more or less completely the feeling and power of moving in all the portions of the body which are supplied with nerves originating from the spinal cord below the part where it has suffered from the external violence. The organ does not recover from this state of inaction as soon as the brain —a day or two at least almost always elapsing before any well-marked sign of improvement is perceptible. *It is probable that the cause of this may be effusion of serum, or blood, occurring in consequence of the injury, which, subsequently undergoing absorption, allows the usual actions to be restored.*"* This great surgeon, however, records no case of concussion lesion of the spinal cord, and none of simple suspension of the functions of the cord by a blow or fall upon the spine.

Passing to works of more recent date we find some interesting facts and cases in a paper "On Injuries of the Spine, including Concussion of the Spinal Cord,"† a paper of especial value in that the testimony, like that of the cases already given, is independent, and wholly unconnected with the class of injuries which usually become the subject of medico-legal inquiry. "Clinical observation," writes the author, "entirely aside from any speculation founded upon physiological and pathological knowledge, has abundantly attested the exceeding gravity of all mechanical lesions of the spine." He records an admirable example of increasing paraplegia and death on the third day, after a "violent injury of the spine" in wrestling, where‡ "the autopsy disclosed a fracture passing through the fifth,

* Italics not in original.
† By John A. Lidell, M.D., Surgeon U.S. Vols. in charge of Stanton General Hospital, Washington, D.C. 'American Journal of the Medical Sciences,' October, 1864, vol. xlviii, p. 305, *et seq.*
‡ Op. cit., p. 306.

sixth, and seventh cervical vertebræ. There was not much displacement, certainly not enough to press on the spinal cord in any way, and not enough to be detected by external examination. The theca vertebralis was lacerated at the place of fracture, accompanied with the effusion of much blood into the spinal canal, compressing the spinal cord. . . . Death had been produced by compression of the spinal cord from this cause. The blood was extravasated from the vessels of the cord itself (arteriæ spinales)."

"Even if the spinal dura mater be not lacerated," he goes on to say, "a quantity of blood may be extravasated from the ruptured thecal vessels between the bony wall of the spinal canal and the theca itself; for its anatomical relations are such that but little force is required to separate it from its osseous case, and in this way the spinal cord may be compressed in a manner analogous to what happens to the brain when blood is extravasated from the middle meningeal artery, between the cranium and the encephalic dura mater." Injuries to the spinal cord itself, we shall do well to remind ourselves, are indeed so dangerous in their results, that when we meet with paraplegia as the result of an injury, inadequate apparently to produce vertebral fracture or dislocation, and the symptoms begin shortly to amend and ultimately to pass away, we are justified in the conclusion that intra-spinal hæmorrhage has almost certainly been the cause of the paralysis by pressure upon the spinal cord. A man slid down a high ladder and landed sooner and harder than he expected upon his buttocks on the ground. When admitted into hospital shortly afterwards he was totally paraplegic. After some weeks he recovered, though whether his recovery was absolutely perfect we cannot say. Another man was thrown out of a dog-cart and shortly afterwards was likewise taken into hospital totally paraplegic. In three days he died, and his cord was found soft and diffluent at the mid-dorsal region. There was no evidence whatever during life of injury to the vertebral column, no sign of fracture or displacement, yet both were found; and on the post-mortem table the displacement could readily be produced, and the cord be squeezed as it had been at the moment of the accident. The second man, in our judgment, died because his cord was crushed; the first man lived because his cord had escaped direct injury.

Here, however, we are more concerned with *concussion* lesions of the *spinal cord*. *"It happens," Dr Lidell writes, "not unfrequently that a paralysis, more or less complete, especially of the lower extremities, is produced by injury of the spine without the occurrence of fracture, or, indeed, of any perceptible lesion of the spinal column or of the spinal marrow. The term concussion of the spinal cord has been employed to designate these cases, because of the analogy they are supposed to bear to concussion of the brain. In both alike, a more or less complete arrest of special function is produced, without any visible injury to the nerve tissue. *Cerebral concussion produces* a state of more or less profound *unconsciousness*, and *spinal concussion occasions* a more or less complete *paralysis* of the parts supplied with spinal nerves, the filaments of which either pass through or are given off from the concussed tract." He then quotes the following case from the 'Medical and Surgical History of the British Army in the Crimea,' which " will afford a clearer view," he says, "of what is understood by concussion of the spinal cord." A private was wounded severely by a shell, and on admission to the hospital "there was a wound about one and a half inches long situated to the right side of the fourth lumbar vertebra. The finger could be passed from it across the spine towards the left side, and a probe passed readily in that direction to the depth of eight inches at least, and seemed to indicate that a foreign body was lodged in that situation. The hips did not correspond in

* Op. cit., p. 321.

shape, and the lower extremities were paraplegic." The wound suppurated, and " the case progressed favorably till the 13th day, when increased inflammation and suppuration, with persistent paraplegia, appeared to demand more energetic treatment than had yet been adopted." Free incisions were made, and an attempt, unsuccessful in its result, to extract a " portion of bayonet supposed to have lodged deep in the hip." The man died on the 21st day, and "on post-mortem examination . . . the posterior portion of the right ilium was found greatly shattered, and the sacro-iliac synchondrosis was completely separated.

. . . *No fracture of the vertebra existed,* nor *were any appearances found in the spinal column sufficient to account for the persistent paraplegia.*"* And hereupon is this somewhat remarkable comment :† " Since concussion of the spine‡ is *per se* but seldom fatal, and therefore the opportunities of investigating its pathological anatomy are much restricted, it has seemed to the writer that no apology would be necessary for quoting the foregoing case in detail." Let us note, however, the severity and widespread nature of the injury, and the immediate paraplegia; and although no appearances were found in the spinal cord, nor was there any fracture of the vertebræ, it is yet open for us to doubt whether in this vast lesion and inflammation the nerves of the limbs may not have been involved, and yet the spinal cord, which ends considerably above the plane of injury, may have itself escaped. Is it not possible, we would ask, that amidst the probable difficulties of post-mortem examinations in the Crimea the nerves outside the spinal column escaped investigation ? Might not the nerves to the lower limbs have been every one torn through ? It stretches the analogy between concussion of the brain and concussion of the spinal cord rather far to assume that

---

* Italics not in the original.
† P. 322.
‡ Why " concussion of the *spine ?*"

a "persistent paraplegia" could have existed and there be yet no lesion discoverable—if carefully sought for—to account for the symptoms. To our mind it seems more reasonable and more in harmony with clinical facts and pathological experience to think that the lesion, which undoubtedly lay behind the *persistent* paraplegia, was not found, than that no lesion was there. We italicise the word *persistent*, for simple *concussion* of the brain may give rise to a transient unconsciousness; and if the analogy holds good, concussion of the spinal cord should *per se* produce a transient paraplegia. We know of no case, nor can we discover the history of any case where this has happened.

An ecchymosed condition of the brain and spinal cord is described by surgical authors as affecting these organs, and Dr Lidell goes on to say: "It cannot be doubted that in at least some instances this ecchymosis, this extravasation of blood beneath the visceral arachnoid membrane into the meshes of the pia mater (connective tissue), denotes a genuine *contusion* of the brain or spinal cord, as the case may be; and that in this way a positive pathological lesion, perceptible to the unaided vision, is *superadded to the concussion.*" But, he rightly says, "*contusion* is much more likely to happen to the brain than to the spinal cord, because, in the *first place*, the surface of the brain is in close relation with its firm unyielding osseous case, while the spinal cord is separated from its osseous envelope by a considerable space, occupied by the cerebro-spinal fluid; and because, in the *second place*, the brain is a more vascular organ than the spinal cord; or, in other words, the brain is much more abundantly supplied with blood-vessels liable to be ruptured by any contusing force than the spinal cord."*

After alluding to the scanty notice taken of "concussion of the spinal cord" by writers on military surgery, of

* Op. cit., p. 322, *et seq.* Italics our own.

whom he names Ballingall, Macleod, Williamson, Hennen, and Guthrie, the author proceeds to give cases met with by himself in military practice. Two of them more especially deserve our attention. The first is entitled "Concussion of the spine from fire-arms, with fracture of the spinous process of the second lumbar vertebra." A lad of 18 was "hit about two inches to the left of the second lumbar vertebra (middle line) by a Minié ball, which passed transversely to the right, and somewhat obliquely forward, through the lumbar muscles, and escaped about five inches to the right of the median line. He fell down immediately, and on attempting to get up again found that his lower limbs were paralysed. The paralysis was complete, both as to sensation and motion, and did not begin to disappear till after the lapse of five days. The bladder was also paralysed." After being taken into hospital several fragments of the spinous process were removed from the second lumbar vertebra, "but there did not appear to be any injury of the lamina, or any other portion of the vertebra." "He complained of pain referred to the right hip, and the paralysis was most marked in that locality." Later he complained "of a queer benumbed sensation which is confined to the right hip and to the right thigh." Two months after the accident, when the last note was made, "his lower extremities and especially the right hip were still weak, but improving, and he now walks pretty well with the aid of a cane only." "This was a case," remarks the writer, "of gunshot wound of the loins with fracture of the spinous process of the second lumbar vertebra, and concussion of the spinal cord." He invites attention to it, "because concussion of the cord is not commonly seen in connection with fracture of a spinous process in military practice. The writer has examined a considerable number of cases of fracture of that process, and the foregoing is the only one among them attended with well-marked concussion of the spinal cord. There is no reason to believe that any other injury was sustained by the second lumbar vertebra besides the fracture of the spine." But who that ponders for a moment on the severity of the blow of a Minié ball, and on the unexpected damage which a bullet may inflict in its passage through the body, can doubt that there was some very definite lesion to account for the symptoms, and that " concussion of the spinal cord " is an altogether inappropriate term to apply to a lesion which the patient's recovery rendered it impossible to discover, and to which the paraplegia was due? The early improvement would seem to point to intraspinal hæmorrhage.

The second case is even more striking, and it seems to us still less deserving of the title accorded to it, " Gunshot wound and concussion of the spine." A middle-aged man was wounded on Dec. 11th, 1862, by a " conical musket ball, which penetrated the lumbar muscles a little to the right side of the spine and produced paraplegia. On exploring the wound with my finger," continues the author, "I found that the bullet had passed forwards, somewhat downwards and slightly inwards, exposing to some extent the body and the transverse process of the second lumbar vertebra. After careful examination I failed to detect any fracture and did not find the bullet. The paralysis of the lower extremities was complete, both as to sensation and motion; the urinary bladder was also paralysed." During January, 1863, he began to recover the power of moving his legs in bed; in February the bladder also recovered its tone, and " during this month the paraplegia continued slowly to diminish. The bullet, a conical one, came away in the dressings." The man continued slowly to improve; in a year he could walk feebly with crutches; in April, 1864, he could walk with a stick, and when the last note was made, in May, 1864, he was " still improving slowly." The same criticism applies to this case as to the other, and, indeed, it seems even more likely that intraspinal hæmorrhage, or damage to some cords of the cauda equina, was the real cause of the symptoms. The nature of the injury, moreover, renders it highly probable that the damage to the vertebral column was more extensive than could be detected by the finger. It is wholly inadequate we think to describe such cases as concussion of the spinal cord, or concussion of the spine, when every circumstance points to the existence of some very tangible lesion, "super-added" though it be to "concussion," a word which should be used to indicate rather the manner of the injury than the result of the injury inflicted by the blow. It should, however, be especially noted that these cases occurred in military practice, and it is not inconceivable that a blow inflicted on a small area with the prodigious momentum of a bullet may be more likely to cause structural damage to the spinal cord or other contents of the spine by "concussion" than is any other injury which befalls the vertebral column.

This conclusion seems borne out by a case recently published by Dr Edmunds ('Brain,' April, 1884, vol. vii, p. 103) of "Concussion and inflammation of spinal cord from gunshot wound of back." The projectile, apparently a large slug, entered obliquely near the inferior angle of the right scapula, and made a tortuous course towards the dorsal vertebræ. There was complete paralysis below the seat of injury. Death occurred four and a half months after the injury, when the bullet wound had healed. No damage whatever had been inflicted on the spine, and the theca vertebralis was intact, but the cord was much atrophied and softened about the level of the wound. Microscopical examination revealed universal softening and myelitis for about two inches opposite the wound, passing gradually into sclerosis of the lateral and anterior pyramidal tracts below, and above into sclerosis of the posterior median columns. The surface of the cord was uninjured, and there was no indication of hæmorrhage either external to or into its substance. Dr Edmunds points out that the myelitis found in this case could not have accounted for the immediate paraplegia, and that the term "concussion" is used to indicate the cause of this.

II

But before going further let us look more closely into the analogy* which exists or which is supposed to exist between concussion of the brain and concussion of the spinal cord, for there can be little doubt, we think, that the term "concussion of the spinal cord" has been derived from the term "concussion of the brain." When a man is stunned by a blow or fall upon his head, common parlance ascribes his state to concussion of the brain. The insensibility may be slight and momentary, or it may be lasting and profound; death may follow almost instantaneously upon the blow, or may not occur until after some hours or days. Is the term concussion of the brain applicable to all these varying conditions which may follow blows upon the head? In cases of transient coma, where recovery takes place without symptoms of injury having been inflicted on the structure of the brain, it is often assumed that some molecular disturbance has been occasioned in the cerebral mass whereby its function has been for the time annulled. Were we to be content with this explanation of the prominent symptom coma, no term could be more suitable perhaps than that which is commonly in use; but we shall do well not to forget what the actual symptoms are, and whereon they depend. The symptoms of cerebral concussion, pure and simple, are essentially those of cerebral paresis of the heart, and they differ in degree only, not in kind, from those of paresis of

* Bearing in mind with Darwin that "analogy may be a deceitful guide."

the heart produced by other injuries which occasion collapse or shock. That the symptoms should be more pronounced after, and be more easily produced by, a blow communicated to the brain than by a blow upon other parts, is neither more nor less than we should expect when we consider what the cerebral mass is, and what are its varied functions. The shake of the cerebral mass may of itself perhaps produce unconsciousness, but that unconsciousness is deepened and maintained by the paresis of the heart and circulation, which is a direct result of the "brain concussion." "Concussion of the brain" as an accurate clinical phrase should be limited entirely, we think, to those cases in which, to quote the definition of Mr Jonathan Hutchinson, there has been a "shake of the cranial contents without any structural lesions of importance."*

We meet, however, with yet other cases in which, while the most prominent early symptoms are essentially those of "concussion"—deep coma and collapse—there is every reason to believe that the brain substance has itself been lacerated or contused. And with reference thereto it is not surprising that there should be differences of opinion as to the share which the actual brain lesion has in causing the symptoms, or even death, in the severer cases of "concussion" injury to the brain. Sir Prescott Hewett says it "still remains to be demonstrated that concussion may prove fatal without leaving a trace of injury to the brain-substance."† He cannot find "a single instance in which the evidence of instantaneous death from simple concussion of the brain will stand the test of anything approaching to a rigid scrutiny." "In every case," he writes, "in which I have seen death occur shortly after, and in consequence of an injury to the head, I have invariably found ample evidence of the damage done to the cranial contents."

* 'Illustrations of Clinical Surgery,' vol. i, p. 86.
† 'Holmes' System of Surgery,' second edition, vol. ii, p. 301.

Intermediate, however, between the cases rapidly fatal, where it is certainly most exceptional not to find injury to brain substance, minute ecchymoses, well-marked contusion or hæmorrhage, and those cases where there has been only transient coma, there are others in which the coma has been longer lasting, where the rallying has been slow, and where the patient may suffer for a length of time from pain in the head, and may show signs of impaired mental power.* Are these symptoms due, he asks, to simple concussion? May they not rather be due to "some extravasation of blood, or to some local injury done to the brain substance?"† Doubtless in many cases which recover without any definite symptoms of motor or sensory paralysis there has been injury to the brain and the lesion is gradually repaired. The opportunities of examining these cases post-mortem are of course few and far between. Prescott Hewett, however, records two ‡ "in which symptoms of concussion of the slightest nature had altogether passed off within a very short time," and in which "well-marked traces of injury were found after death in the brain." "In one case, in which after a blow on the head, there had been mere giddiness for a few minutes and then complete recovery, some patches of contusion were found at the base of the brain; marked by minute specks of blood closely clustered together, these patches were, in two or three places, of the size of a shilling, and extended, about a line in depth, into the structure of the brain: there were no disseminated specks of extravasated blood." In the other case, where the symptoms of

* Since the above was written we have had a case in hospital where there was every reason to believe from the continued severe pain that the præfrontal region of the brain had been contused. The patient had fallen and struck the back of his head. He recovered after about three weeks' rest in bed. There were never any one-sided symptoms, nor elevation of temperature, nor changes in the optic fundus.

† Op. cit., p. 307.   ‡ Op. cit., p. 303.

concussion soon passed off and the patient died of some other disease eight days after the accident, "thin layers of extravasated blood" were found in the cavity of and adhering to the parietal layer of the arachnoid, and "in the centrum ovale, close to the right side of the corpus callosum, and extending partly into it, was an extravasation of blood of the size of a nut."

That patients may recover after severe injuries to brain substance received in so-called simple concussion, is evidenced nowhere better than by another case where Sir Prescott Hewett was able to examine the brain of a man who, twenty years before, had been under his care for a very severe injury of the head marked by "the symptoms of so-called concussion of the brain, but there was no sign of fracture of any part of the skull. For several days he struggled between life and death, in a state of perfect unconsciousness, followed by a violent delirium, which ultimately, however, subsided, and in a few weeks he was so far well that he was able to leave the hospital."* He resumed his occupation, and was an able workman, clear in intellect as before, and he did not suffer more from headaches than other people. The autopsy revealed extensive excavations in the anterior parts of both hemispheres of the brain, filled by loose areolar tissue and serum. That of the right side was about twice as large as the excavation in the left, and measured about an inch and a half in diameter, with a depth of about an inch, "and so placed that the inferior margin lay close to the base of the brain, whilst the inner one was close to the median fissure."†

* Op. cit., p. 320.

† An example of that which Dr Ferrier, both by clinical and experimental observation, has so clearly established ('Localisation of Cerebral Disease,' p. 33), "that sudden and extensive lacerations may be made in the præfrontal region, and large portions of the brain-substance may be lost without causing impairment either of sensation or motion; and, indeed, without very evident disturbance of any kind, bodily or mental, especially if the lesion be unilateral."

Although then there are "no characteristic signs by which contusion of the brain can be clearly recognised," we are justified, it seems to us, in agreeing with Hewett that the brain substance has probably been bruised whenever "the symptoms are severe after an injury of the head;" but the cases which we have quoted seem to show this also that "concussion symptoms"—coma and collapse—are not necessarily due to brain lesion, for the brain lesion in all remained although the "concussion symptoms" passed away.

That there may be an entire absence of symptoms due to slight contusion of the brain *per se* in such cases of concussion would seem also to be the opinion of Mr Hutchinson, from whom we have already quoted, and whose teaching is of the highest value on all questions connected with injuries of the head. Plate XVI of his 'Illustrations of Clinical Surgery' gives a beautiful delineation of contusion of the surface of the brain, such as is often met with in cases where the patient dies directly from the effects of concussion. At the autopsy* "certain prominent parts of the convolutions, more particularly of the anterior lobes close to the extremities of the olfactory bulbs, and of the lowest parts of the middle lobes, show slight surface contusion. At most places a little ecchymosis into the pia mater is all that can be proved, but just in front of the ends of the olfactory bulbs there is more than this, and the tips of the bulbs themselves and the convolutions in front of them have evidently been somewhat broken by contusion." "Lines of fracture in the base of the skull were found, and a thin clot of blood in the arachnoid cavity," and "yet," remarks Mr Hutchinson, "neither these lesions nor any of those depicted in the sketch were in themselves sufficient to have caused death, nor even probably to produce alarming symptoms." . . . "Taken collectively these fractures and contusions were proof that the head had been very violently concussed, and it was from the effect of such concussion

* Op. cit., p. 83.

upon the cerebral mass that the man died, and not from any one of the lesions mentioned." . . . "This distinction," he goes on to say, "is I think very important. It would be easy to put down such a case as death from fracture of the base of the skull or death from contusion of the brain, for both these conditions were undoubtedly present, but the symptoms are, I think, conclusive that they were merely the concomitants of the fatal injury. I cannot but believe that many cases of concussion which recover are attended by surface lesions at least as extensive as those seen in this sketch. There is nothing in them *per se* which would be in the least likely to cause death." And again he expresses his "conviction that a considerable number of the head cases, fatal within periods of a few hours or a day or two, die from the general effects of the shake of the cerebral mass. Lesions are found, it is true, but they are to be regarded, I must repeat, as indications of the violence of the shake and not as causes of death, nor perhaps even as serious complications."

Whether, then, we incline to the view that the brain lesions are the chief cause of death in such cases, or to that which attributes death to shake of the cerebral mass, independently of the specific lesions themselves, there is we see no difference of opinion as to lesions being well-nigh invariably met with in cases of severe fatal concussion, nor as to the probability of there being definite lesions, surface contusions, and ecchymosis of the brain substance, in many cases where there have been symptoms of severe "concussion" even though not proceeding to a fatal issue.

We may fitly divide cases of "concussion" injuries of the brain into three classes.

A. Cases of "concussion" where the symptoms of concussion are essentially transient, momentary only or lasting for a few minutes, and are mainly due to the sudden "shock" induced by shake of the whole brain mass by a blow upon the head.

B. Cases where the early symptoms of "concussion" proper are of longer duration, and the later—pain, irritability,* &c.—are slow to pass away, and where there may not be, although undoubtedly there very often are, definite structural lesions of the brain substance at a point remote it may be from the part struck, lesions by *contrecoup*, which of themselves may give rise to no symptoms.

C. Cases rapidly fatal with concussion symptoms, or fatal at a later time after the symptoms of concussion have materially subsided, where brain lesions are almost invariably found after death, with or without damage to the membranes and the skull, and the influence of which lesions in causing death is differently estimated by writers on surgery.

This is we trust a fair and comprehensive classification of the various cases of so-called "concussion of the brain," which are not followed by secondary symptoms such as are due to inflammation and death of the cranial bones, or to inflammation and suppuration of the brain and its membranes. We have seen that there are strong grounds for the belief that "concussion" of the brain, even when the "concussion" has been only slight, is very frequently associated with structural lesions of the brain itself, lesions which of themselves may give rise to no separate symptoms, and that the usual symptoms of concussion, varying greatly in degree, are more those of cerebral paresis of the heart, and of the bodily functions generally. *Shock* or collapse, differing in degree only from shock following injuries elsewhere, is indeed the most prominent evidence of brain concussion, and common to that shock and to all true shock however caused is this one symptom most pronounced, mental enfeeblement or annihilation of consciousness for the time being.†

* See Mr Hutchinson's table of the after-symptoms quoted on pp. 186, 187.

† Thus, Mr Spence writes ('Surgery,' second edition, vol. ii, pp.

What is the organ, however, and how is it situated, the concussion and violent shake of which give rise to this train of symptoms? The cerebral mass, a large soft solid, surrounded by envelopes or membranes, fitting closely to and filling up the crevices and grooves in the skull which is its protection and covering—an organ so built and placed that it must be liable to commotion from any blow of sufficient force upon its osseous case. We need hardly inquire into the physical causes of this liability. Is not the common experience of surgeons in all countries adequate of itself to establish a belief in the truth of the exceeding liability of the cerebral mass to suffer in the way described from the effects of a blow, whether that blow be direct or indirect upon the cranial walls?

But when we leave the cranium and enter the spinal canal what happens? " In leaving the skull the dura mater is intimately attached to the margin of the foramen magnum; but within the vertebral canal it forms a loose sheath around the cord (*theca*), and is not adherent to the bones, which have an independent periosteum. Towards the lower end of the canal, a few fibrous slips proceed from the outer surface of the dura mater to be fixed to the vertebræ. The space intervening between the wall of the canal and the dura mater is occupied by loose fat, by watery areolar tissue, and by a plexus of spinal veins."*
. . . . " On the *spinal* cord the pia mater has a very different structure from that which it presents on the

710, 711), "The symptoms seem dependent on some temporary derangement of the brain substance, or its circulation, giving rise to functional disturbance of the sensorium, which gradually passes off *with the subsidence of the shock which caused it.* . . . . In the very rare cases where concussion terminates fatally without reaction occurring, and in which no obvious lesion is found, it is probable that the *shock* has affected not only the sensorium, but also that portion of the nerve-centres more essentially concerned in organic life." Italics not in original.

* 'Quain's Anatomy,' 7th edition, vol. ii, p. 563.

encephalon, so that it has even been described by some as a different membrane under the name *neurilemma of the cord*. It is thicker, firmer, less vascular, and more adherent to the subjacent nervous matter: its great strength is owing to its containing fibrous tissue, which is arranged in longitudinal shining bundles."* .... "At the *base* of the brain and in the *spinal canal* there is a wide interval between the arachnoid and the pia mater. In the base of the brain this subarachnoid space extends in front over the pons and the interpeduncular recess as far forwards as the optic nerves, and behind it forms a considerable interval between the cerebellum and the back of the medulla oblongata. In the spinal canal it surrounds the cord, forming a space of considerable extent."† It is in this subarachnoid space where the chief part of the cerebro-spinal fluid is lodged. Thus we not only have the cord surrounded by fluid and areolar tissue, but we also find it carefully separated from, and slung as it were to, the sides of the canal in which it lies. The *ligamentum denticulatum*, together with the off-shooting nerves, forms the special contrivance whereby the cord is kept and fixed in its place, and it consists of a "narrow fibrous band which runs along each side of the spinal cord in the subarachnoid space, between the anterior and posterior roots of the nerves, commencing above at the foramen magnum, and reaching down to the lower pointed end of the cord. By its inner edge this band is connected with the pia mater of the cord, while its outer margin is widely denticulated; and its denticulations, traversing the arachnoid space, with the arachnoid membrane reflected over them, are attached by their points to the inner surface of the dura mater, and thus serve to support the cord along the sides and to maintain it in the middle of the cavity."‡ And as if the spinal cord were not thus sufficiently protected by special attachments and by paddings and buffers, we find it in a bony canal,

* Ibid., p. 564. † Ibid., p. 565. ‡ Ibid., p. 566.

whose walls are of great thickness, many times the thickness of the bony wall which surrounds the brain. If there be any exception to the security which the cord derives from its mode of attachment, it will be found, we are inclined to think, in the cauda equina. Examine the spinal cord lying *in situ* in the recently opened spinal canal, and we shall see that, at the lowest part, where the great mass of nerves is collected to form the cauda equina, say from the eleventh dorsal vertebra downwards, the contents of the spinal canal have much less room, that there is less of padding, less of detachment and separation from the bony wall, and insomuch perhaps there is a greater liability for the cord to suffer at this point from the effects of such a blow as that which, were it upon the cranium, would produce "concussion" of the brain. Yet even here some compensation is to be found in the increasing thickness of the spinous processes of the vertebræ, whereby the spinal contents are still further removed from the surface of the body. Such, then, is the wondrous security of the spinal cord. We need not dwell longer on the physical causes thereof. Let us appeal to the common experience of surgeons in all countries and ask, Is it not the fact that injuries to the spinal cord from simple blows upon the spinal column, from such blows as would cause " concussion of the brain," are amongst the very rarest to which the human body is liable? And it would be lamentable indeed were it not so: for carrying, as we have before remarked, its whole function as a conductor in each horizontal segment of no matter how microscopic thinness, not only *must* the spinal cord be thus securely protected, but by its very constitution also is it compelled to tell and warn us —as infallibly it does tell and warn us—when injury has been inflicted upon it.

Let us, however, seek for cases which may be placed in divisions analogous to the three divisions of concussion injuries of the brain.

With Class A, cases of "concussion where the symptoms of concussion are essentially transient, momentary only, or lasting for a few minutes, and are mainly due to the sudden 'shock' induced by shake of the whole brain mass by a blow upon the head," we should place examples of transient paraplegia after blows upon the spine, where the concussion of the blow having been communicated to the spinal cord, that portion of the cord supposed to be concussed is rendered temporarily *hors de combat*, ceases for the time being to discharge its habitual function, and then, the effects of the concussion, shock, or shake having died away, the habitual function is restored, and there are no subsequent ill effects. If there be such cases, then without doubt the analogy holds good. It is a remarkable fact, however, that nowhere in medical or surgical literature have we been able to find the record of even one such case. The experience of surgeons, even of those who have had the greatest opportunity of seeing serious accidents is, we take it, pretty unanimous that cases of the kind are excessively rare, and, although we have made careful inquiry upon the subject, we have been unable to meet with any case which would stand the test of rigid scrutiny. Dr Wilks, however, writes ('Diseases of the Nervous System,' Edit. 1st, p. 201), "I have more than once seen a man receive a severe injury to the back and be taken up paralysed, but in a few days he has perfectly recovered the use of his limbs, just as in concussion of the brain with loss of consciousness and rapid recovery; there was, in fact, a stunning of the cord."

We are not dealing, be it remembered, with cases of general motor and sensory enfeeblement in *severe "shock"* after falls and like accidents, where it might be said that the whole cerebro-spinal system had been concussed, and that some of the symptoms were due thereto. What we are rather looking for is a case of *transient* paraplegia, or annihilation of the function of the spinal cord, by a blow upon the spine communicated to the spinal marrow.

The following case is recorded by Abercrombie (op. cit., p. 373), who writes :—" In summer 1816, I saw a man who had been employed in blowing a rock near Edinburgh. Not having retired to a sufficient distance, and standing with his back to the rock when the explosion took place, a large piece of stone struck him on the spine about the lower dorsal and upper lumbar vertebræ. He instantly fell, completely deprived of power in the lower extremities. I found him in this state a few hours after the accident, when he also complained of violent pain, beginning in the seat of the injury, and extending downwards along the thighs. On the back there was an extensive swelling, which made it impossible to ascertain the state of the vertebræ. He was confined to bed for several weeks without any power of his lower extremities, and with considerable difficulty in passing his urine, but gradually recovered, and in a few weeks more, was free from complaints."

This was regarded by the author as an instance of true concussion of the cord, but it appears open to very grave question whether the symptoms were not in this case also attributable to hæmorrhage. We are told that the man instantly fell, completely deprived of power in the lower extremities, but we know nothing as to the existence of collapse, which in all probability existed, and during which there may have been ample time for an extravasation of blood sufficient to cause paraplegia. The clinical history looks much more like a case of intra-spinal hæmorrhage which was gradually absorbed, although the nature and force of the blow were those of a bullet or shell, and the spinal cord may really have been injured by concussion as in Dr Edmunds' case. The history of the two cases is, however, markedly different, and there is, moreover, but scanty evidence as yet to show that a complete disorganisation of the cord, such as on the concussion theory must have been present to account for the symptoms, can ever be repaired.

It has, however, been suggested by Mr Savory that the abolition of motor reflex action in cases of fracture-dislocation when there has been contusion of the cord, is an evidence of *concussion* of the cord, not at the part alone where there has been contusion, but of the whole cord beyond the seat of lesion. He points out that where the cord has been experimentally divided in animals the motor reflex function of the cord remains when the shock of the operation has passed off ("Notes on Concussion of the Spinal Cord and Brain," 'Lancet,' vol. ii, 1882, p. 883), but that in man it is abolished after fracture. And putting aside the local lesion as inadequate to explain it he asks, " Would it not be more correct to say that the violence injured the cord beyond by concussion ? " This view receives the support, moreover, of Dr Lidell, from whom we have already quoted, and whose various writings show how great is the loss that surgery has to deplore in his recent death, in an! important and most valuable paper " On Contusions of the Brain and of the Spinal Cord " ('American Journal of the Medical Sciences,' July, 1883). He records a case (at p. 56) which is thus summarised:— " Contusion of the spinal cord caused by a blow on the back from a falling tree; paraplegia; death six days after the accident. Autopsy: linear fracture of the first and second dorsal vertebræ (*i.e.* without displacement) also present." In addition to this the anterior and posterior ligaments were somewhat lacerated or detached, and there was some hæmorrhage, not enough to cause pressure on the cord, external to the theca vertebralis. The usual appearances on section of the cord at the seat of lesion are minutely described. They were restricted to the locality of the vertebral fracture, and were symmetrical in both lateral halves of the cord. Dr Lidell then goes on to say that the patient suffered from "concussion as well as from contusion of the spinal cord, and consequently exhibited reflex motor paralysis." The local contusion suppressed the function of the cord as a conductor of impressions up and down, "the concussion of the cord suppressed its functions as a series of independent nervous centres arranged one above another, and in this way produced reflex motor paralysis of wide extent." This is in fact the theory of Mr Savory applied to a particular case, but it cannot escape observation that the concussion fell only on the parts of the cord below the seat of local lesion, and not on those above it also. Mr Savory has noticed this objection, for he writes: "In any case it is not clear why the part above does not to a greater extent lose its reflex power," at any rate almost though not quite invariably. It seems to us that the right explanation of the phenomena probably lies in the fact that the cord, in man at least, is not a series of "independent nervous centres arranged one above another," but that each lower centre is more or less dependent for its full and natural functional activity on the perfect integrity of the parts above it. Dr Hughlings Jackson's views with reference to the evolution and dissolution of the nervous system apply to the whole central nervous system, to the spinal cord as well as to the brain, and the phenomena which follow contusion of the cord seem to support them. Is there not enough in local contusion to explain the symptoms without calling in concussion of a wide region of the cord, a region, moreover, which admittedly received no structural injury? For Lidell says: "No doubt therefore exists that concussion of the spinal cord *per se* is not attended with any change in structure or appearance which is at present recognisable by the anatomist after death." Why should concussion have only fallen on the cord between the level of the lesion and the filum terminale, rather than between the lesion and the medulla oblongata? Is it conceivable that the concussion could have told in one direction and not in the other? Is it not more likely that the real explanation of the phenomena lies in the cord itself, and that there is a gap yet to be filled up in our knowledge of the physiology of the cord, and that we cannot as yet precisely account for the fact that a local lesion annihilates a function which appears to have an independent seat in the parts of the cord beyond it? Why, moreover, should this reflex motor activity be sometimes increased or exaggerated by local lesion—by a myelitis, for instance, at a certain level—be increased below the lesion but not above it?

It would be idle, perhaps, to contemplate the abolition of a phrase "concussion of the spinal cord," which has been sanctioned by such men as Astley Cooper, Abercrombie, Mayo, Brodie, and others, but when we use it let us be quite clear as to what we mean, and let us know what are and what are not the cases of injury to the spinal cord to which the term is strictly applicable, and which bear analogy to those with which we are familiar under the term "concussion of the brain."

But, secondly, we have to look for cases analogous to those of concussion of the brain in which the "early symptoms of concussion proper are of longer duration, and the later—pain, irritability, &c.—are slow to pass away; and where there may not be, although undoubtedly there very often are, structural lesions of the brain substance at a point it may be remote from the part struck, lesions by *contrecoup*, which of themselves may give rise to no symptoms" (p. 26 *ante*). We are met at once with this difference between injuries to the brain and injuries to the spinal cord, that structural lesions of the cord, even small in size, invariably give rise, sooner or later, to recognisable symptoms. We may go further and say that if these symptoms are of long duration, there has indubitably been some lesion either to the spinal marrow itself, or that its function has been impaired by pressure upon it from without. It is well recognised, we think, that dislocation of the vertebræ, either with or without fracture, may occur without giving rise to any perceptible deformity, the bones having sprung back by the resilience of the ligaments into their natural positions. The records of the dead-house leave no doubt on this point; and over and over again the spinal cord has been found crushed by displacement of vertebræ, when there has been no physical sign whatever during life of injury to the spinal column. This has been seen so often that when that marked symptom, paralysis of motion and sensation, follows a fall, we are justified in concluding either that the cord has actually been crushed, or that, if the displacement has not been sufficient to disorganise the cord, as in those rare cases where recovery ensues, it has yet been enough to lacerate some of the blood-vessels which surround the marrow, and the function thereof has been for the time annihilated by the pressure of extravasated blood. At any rate it is essential to exclude these and other kindred explanations before we admit " concussion " by itself as the true explanation of any traumatic lesion of the substance of the spinal cord. A very interesting case is recorded by Mr Hutchinson, which bears on this point.

*A sailor was violently dashed by a wave in a storm against a skylight, striking it, he thought, with his head. He was insensible for five minutes. He had intense pain in the back of the neck, and from the first the lower limbs felt numb and useless. The arms also were partially paralysed. His mouth was slightly drawn to one side. When admitted into hospital three months after the accident he was able to move his legs, but could not stand without supporting himself with his hands. The bowels were constipated, but he had power over the bladder. Sensation was more perfect in the left leg than in the right, and he thought his left arm and leg were stronger than the right.

* 'Medical Times and Gazette,' vol. i, 1879, p. 348.

He could not move his neck so freely as before the accident. Galvanic currents to the lower limbs began to give relief at once, and he left the hospital in two months, able to walk without support and improving daily. In his remarks upon this case, Mr Hutchinson said he was inclined to think that "the paraplegia having been almost coincident with the injury, dislocation of one of the vertebræ might have occurred, causing injury to the cord. He had seen several cases where such dislocation had occurred, and where the vertebræ had easily returned into their places; so that it was only upon very careful post-mortem examination that the diagnosis could be verified."

Even in such a case* as that recorded by Mr Shaw, where instant paralysis of both upper and lower extremities was the consequence of a fall on the head and shoulders, and where, after death in sixty hours, there was found neither fracture nor dislocation, nor blood effused in the canal; "but on making a section of the cord opposite the third and fourth cervical vertebræ, a clot of blood was found lying in its centre," we may fairly, it seems to us, agree with the author in thinking that the lesion in the cord may be satisfactorily accounted for otherwise than by the phenomenon of concussion. "It does not appear improbable," he writes, "that a sudden, violent and extreme bend or twist of the flexible cervical region of the spine should have been caused by a fall headlong from a height; or that, with that excessive bending of the spine, the cervical enlargement of the cord should have been bent likewise; or that in that sharp, acute, abrupt flexion of the cord, a blood-vessel should have been ruptured in its interior, and a clot of blood be deposited." There is a limit to the bending capacity of the spinal cord as well as of the spine, and we strongly suspect that in many cases of fracture and dislocation of the spine, or even where the spine has been uninjured, extravasation of blood within the cord, and dis-

* 'Holmes' System of Surgery,' 2nd edit., vol. ii, p. 371.

integration of the structure of the cord, have been caused by severe and sudden bend. Spinal cords no doubt differ in this respect, but there can be no question that every possibility of a severe bend should be excluded before a case is accepted as one of concussion lesion of the cord.

In the most valuable series of " Cases of Paraplegia," published by Sir William Gull in the 'Guy's Hospital Reports' for 1856 and 1858 (series iii, vols. ii and iv), is an example of the same kind of injury. It is entitled (vol. iv, p. 189) " Paraplegia supervening two days after a violent exertion in lifting a heavy weight; softening of the cord opposite the fifth and sixth dorsal vertebræ, no injury of the membranes, ligaments, or bones of the spine. Death after six weeks." A man, aged 25, of a rather delicate constitution, felt a sudden pain in his back after lifting some heavy deals. On the morning of the second day he found on waking that his legs were paralysed, and when admitted into hospital on the fourth day there was complete paraplegia, his urine was ammoniacal, and a bedsore had already begun to form over the sacrum. No trace of injury was found after death in the ligaments or bones of the spine, but "opposite the fifth and sixth dorsal vertebræ the cord was softened through all the columns into a thick, greenish muco-puriform fluid with a tinge of brown," and under the microscope it was "seen to consist of disintegrated nerve-tissue with a few irregular collections of granules." " There was no evidence of any plastic exudation."

In another most remarkable case recorded by the same author (Case xxix, vol. iv, p. 200) the usual immunity of the cord from injury in such accidents as the two last cases exemplify, is shown very strikingly by the damage which the cord received in an otherwise trivial accident, because of the existence at the time of meningeal disease. The patient was a nurse in Guy's who had been under observation for some years in consequence of anæsthesia, partial paralysis, and wasting of both arms, symptoms which had had no known cause for their beginning. One day in December she " accidentally fell forwards upon the stone steps of the hospital, from stepping upon her dress whilst assisting a patient into a cab. Her left temple was cut, and she was rendered insensible by the fall. On recovering consciousness a short time afterwards the legs were found to be quite paralysed, and there was almost entire loss of sensation." There was the usual train of symptoms, ammoniacal urine, sloughing, &c., and in a month she died. Extensive disease was found in the cord and its membranes. "In the dorsal region there were plates of true bone, formed by ossific degeneration of the inner layers of the thickened dura mater. . . . They merely enveloped the cord without producing any pressure upon it." . . . " In the dorsal region the anterior columns were ruptured transversely across, apparently at a recent date, and probably by the fall which brought on the fatal symptoms." And commenting on the case the distinguished author writes : "The fatal accident was peculiar. The adhesions of the membranes prevented the movements of the cord in the sheath, and exposed it to stretching by any sudden motion of the spine." The form of accident which in this case was adequate to cause so grave an injury to a spinal cord already affected with disease, may doubtless cause the same kind of injury to a perfectly healthy cord when subjected to sudden, violent, and extreme, bending of the spinal column as in the case recorded by Mr Shaw. But we should err in calling such cases " concussion of the spinal cord."*

* Here it may be remarked that we do not cease to call a case "concussion of the brain" because it is associated with cerebral hæmorrhage. At first truly not, but later, when the symptoms of shock have subsided, and there is paralysis here or there, we change the name, and no longer attribute the symptoms to concussion, but speak rather of "fracture of the base," "fracture of the skull," "cerebral laceration," "hæmorrhage," as the case may be.

Let it be noted, however, that these patients died; and here we shall do well to remind ourselves how exceedingly dangerous are lesions of the spinal marrow, how prone they are to run on to inflammation, and by a gradually extending myelitis to destroy the functions of the cord in those parts which are necessary to life, and thereby lead to a fatal issue. How rare it is to see a man recover after injury to his spinal cord in cases of fracture or dislocation of the vertebral column. The length of time he lives may vary, and does vary according to the site of the lesion, but how commonly fatal these injuries are, and how few surgeons have ever seen a case of recovery after such undoubted injury to the cord.* The exceptions quoted are exceptions which emphasise the rule, and they have been recorded for this very reason that recovery is so wholly unusual a result. The man dies, not because the bones of his back are broken, or because one vertebra has been separated from another, he dies because of inflammation of the spinal cord after it has been crushed, lacerated, or contused. " It is the progress of inflammation to the spinal marrow," says Sir Charles Bell, "and not the pressure or the extension of it, which makes these cases of subluxation and breach of continuity of the tube fatal."† Nay, is not death so almost invariably the result, that common experience tells us that if we see recovery taking place, and the function of the cord being restored after complete paraplegia, the case is one most likely of hæmorrhage into the vertebral canal which has pressed upon the marrow? And is not the existence of this hæmorrhage, and of hæmorrhage only, in such cases a very present hope to us in time of trouble? How unlikely then it is that we should meet with cases of undoubted lesion of the spinal cord itself produced by con-

* Vide 'Lancet,' vol. i, 1881, p. 460, a case brought before the Clinical Society of London by Berkeley Hill, of fracture of the spine treated with Sayre's jacket.
† 'Institutes of Surgery,' vol. i (1838), p. 152.

cussion blows upon the spine, and find that they too have not an equal liability to run to a fatal end. Are not the probabilities immense that when paraplegia follows a severe blow upon the spinal column, and the paraplegia gradually subsides, and the function of the cord is again restored, such paraplegia has been caused by the pressure of hæmorrhage outside the spinal marrow rather than by injury to the spinal marrow itself?

Examples of paraplegia from blows upon the spine are to be met with in surgical literature. We have already referred to Brodie's case* of paralysis after a severe blow upon the spine where the man regained the use of his limbs in the course of three or four weeks, and we have heard the conjecture which he makes as to the probable causes of the palsy. Even more striking than that, and confirmatory of that which we hold to be the true pathological explanation of Brodie's case, is the case recorded by Mayo.† "A man received a violent blow on the three inferior lumbar vertebræ from a log of wood which fell upon him. He died in four hours. Extravasated blood was found in the spinal canal, but the vertebræ were entire and the cord was healthy." Why, even a blow like this, violent as it in all probability was, did no damage to the important nervous structures lying securely within the vertebral canal. Had the man recovered, might not this case also have been placed on record as one of "concussion of the spinal cord?"

How dangerous to life these injuries are, even when the lesion of the cord is of limited extent, is admirably shown by the following case, recorded by Gull in the paper already named (Case xxiii, vol. iv, p. 191, 1858). A porter, aged 33, strong and healthy, was carrying a sack of coals on his back down some cellar-stairs, when his foot slipped forwards from under him and he fell, the sack of coals falling

* 'Medico-Chir. Transactions,' vol. xx; vide *ante*, p. 5.
† 'Outlines of Pathology,' p. 165.

upon him. There was loss of sensation immediately after the accident, followed by hyperæsthesia, paralysis of legs, left arm and sphincters. He died thirty-four hours after the accident. "*Post-mortem examination.*—The spine only was examined. There was no external trace of the injury; no displacement of the vertebræ discoverable by external examination. The membranes of the cord were healthy. Opposite the fourth and fifth cervical vertebræ the substance of the cord was contused. On section there was found ecchymosis of the posterior horn of grey matter on the left side, and of the adjacent part of the lateral and posterior columns. There were also limited spots of ecchymosis on the right side, one in the right posterior column, and one in the anterior cornua of the gray substance. . . . . On examining the spinal canal after the removal of the cord nothing abnormal was discoverable in the bodies of the vertebræ opposite the lesion of the cord; but on dissecting off the posterior ligament it was seen that the body of the fourth was separated from that of the fifth, and that the left articular process of the fourth had been chipped off by the violent pressure of the lower one against it." The case is an exceedingly important one, and more deserving than are most cases of the title which the author gives to it of "Concussion of the cord in the cervical region from direct violence." At the same time, however, it is well to recognise that there was severe injury to the spinal column, such an injury in fact as might readily have been produced by violent wrench of the flexible cervical spine; and it is not inconceivable that in that wrench the spinal cord was likewise severely bent, and that thus, rather than by "concussion," the spots of ecchymosis were produced in the cord. While, however, we may accept this case, even though with hesitation, as one of concussion lesion of the spinal cord, not seeking to lay undue stress on the objection we have offered, the same title is hardly applicable, we think, to the next case which we quote from the same author (Case xxiv, op. cit., p. 193), " Concussion of the cord by a fall; recovery of power after two hours. Subsequent effusion of blood outside the theca vertebralis in the neck. Paraplegia of upper and lower extremities. Paralysis of intercostals. . . . Death in fifty-five hours." A man, aged 40, fell backwards from a moderate height, a heavy plank falling at the same time upon him. Brought at once to the hospital he was found collapsed but sensible. There was entire paralysis of the left leg, partial of the right, and also partial paralysis of the arms, but he was still able to flex the fingers. After two hours he had so far recovered from the immediate effects of the injury that he could draw up his legs and grasp the hand; the circulation was improved; surface warmer; no injury of spine discoverable. The next morning he was absolutely paralysed in all his limbs. There was loss of sensation in the paralysed parts, and the ribs scarcely moved in inspiration. "*Post-mortem examination by Mr Bryant.*—No external evidence of injury to the spine. On dividing the soft parts there was found a separation between the fourth and fifth cervical spinous processes, and dislocation of the articular processes. The interspinous and capsular ligaments were torn through. Extravasation of blood outside the theca vertebralis on its anterior aspect. The effused blood compressed the cord, which was otherwise uninjured. After careful examination there were not found any signs of bruising of its tissue. The extravasation apparently arose from injury to the lower part of the body of the fourth vertebra, which had been fractured, and the intervertebral substance torn. The calibre of the canal was slightly encroached upon by displacement of the fourth vertebra, but not so as to press on the cord. The extravasation, though most abundant opposite the injury, extended downwards to some distance. The membranes of the cord were uninjured." In giving a title to this case the author has assumed, if we mistake not, that the function of the cord was for the time annihilated by the force of the blow received in the fall, and that the return of power after two hours was an evidence of the truth of this surmise. It must, however, be noted that the return of power was coincident with the subsidence of the collapse, and it is, moreover, remarkable that the paralysis which resulted from the supposed concussion of the cord was only partial. It seems to us almost impossible that a concussion blow, of such violence as it must have been in this case, could have been limited in its effects upon the portion of the cord concussed, and we should find an explanation of the phenomena rather in the irregular distribution of the effused blood, which continuing slowly to increase, at length induced the total paralysis to which the man succumbed. Doubtless the fatal result in these cases was largely due to the injury having been inflicted on the cervical portion of the spinal column, and we should perhaps be justified in expecting that were like injuries inflicted upon lower portions of the cord, the fatal result might be less rapid, or even recovery might ensue.

In vol. viii of the 'Transactions of the Pathological Society,' Mr Curling records a case which was "viewed as one of concussion of the spinal cord." A boy, aged 8, had fallen, as far as could be learned, upon his nates while sliding in the gutter. Five days after the accident he was admitted paraplegic. "After manifesting a remarkable tenacity of life" he died in three and a half months. The cord was found softened in several places.

In the 'London Medical Record' (vol. iv, p. 83) will be found the notice of two cases recorded by Dr Lochner, of Schwalbach ('Aertzliches Intelligenzblatt,' Oct. 19, 1875). The first case was a man, aged 52, almost an imbecile, who on May 23rd, when drunk, was struck by a comrade and fell backwards to the ground, with his back across a stone. He died in a month, and opposite the eighth and ninth dorsal vertebræ the whole cord was softened. Dr

Lochner remarks that he "would still have doubted if the spinal cord could be severely injured whilst the investing canal was still intact, had not a similar case occurred in the practice of his colleague, Dr Bruglocher." A man fell down some stone steps and was picked up dead. Various lesions from *contrecoup* were found in the brain, and the "spinal cord at the level of the sixth cervical vertebra was torn across," so that more than half of it was severed. " It appeared that the neck and head had been forcibly bent backwards in the fall," a circumstance of the accident which seems to us to detract from the merit of this case as an example of " concussion injury." As far, however, as the histories warrant, there seems no reason to withhold this explanation from the rare and fatal cases of Mr Curling and Dr Lochner. And not less remarkable than theirs, is the case recorded by Abercrombie (op. cit., p. 343) as having occurred in the practice of Dr Hunter. " A man, æt. 36, was thrown from the top of a waggon, a height of about ten feet. He alighted upon a pile of small stones in such a manner that the shock on first coming in contact with the top of the pile was received on his back between the shoulders. He attempted immediately to get up, but fell instantly from complete loss of power of the lower extremities; and very soon after he had an involuntary discharge of urine and fæces." After the lapse of a month he was admitted into the Edinburgh Infirmary, and died with tetanic symptoms in about five days.

" *Inspection.*—No injury could be detected in any of the bones of the spine. There was a high degree of vascularity of the pia-mater of the cord, especially at the upper part of the dorsal region. There was most extensive *ramollissement* of the body of the cord, which affected chiefly the anterior columns. These were most remarkably softened throughout the whole course of the cord, in many cases entirely diffluent, and the softening was traced quite to the upper part of the cord, and affected the corpora pyramidalia.

The posterior columns were also softened in many places, though in a much smaller degree, not diffluent like the anterior, but breaking down under very slight pressure."

This case is often regarded as one of uncomplicated concussion injury, but does not the nature of the accident render it highly probable that the original lesion of the cord, which was the starting-point of the extensive myelitis, was caused by a sudden twist or bend?

Recorded examples of *undoubted cord injury which progressed to recovery* are exceedingly rare. We say *undoubted cord injury*, for it is essential to exclude the occurrence of lesion of the membranes or the spinal nerves, structures which, in fracture or dislocation, may be more easily injured than the cord itself by *slight* displacement of the vertebræ, or by limited effusion of blood external to the marrow.

A case of Mr Hutchinson's provides the only example of recovery which we can find. "Paraplegia from myelitis following injury to the back."* A girl, aged 16, the subject of inherited syphilis, was admitted into hospital on Feb. 6th. A fortnight before admission she "slipped down about twenty steps of stairs on her back." She felt no inconvenience whatever for two days beyond pain in her back. On the morning of the fourth day she had "pins and needles" in her legs, and found she could not stand up or sit in an upright position. About the same time she had incontinence of urine and fæces. The motions came away involuntarily for two days, and for about eight days afterwards the bowels were not relieved, the incontinence of urine continuing. The bowels were then slightly opened by purgatives. Entire loss of power over the limbs, the bladder and the bowel, remained until her admission. There was then complete loss of motion and sensation in the lower limbs, and incontinence of urine and fæces. There were some bedsores about the lower part of the back, but no

* 'Medical Times and Gazette,' vol. i, 1879, p. 348.

irregularity indicative of fracture. There was slight pain on pressure along the lower part of the spine. On Feb. 10th she was "gradually recovering power. Much hyperæsthesia. Cannot bear the bed-clothes on the feet. Incontinence of urine and fæces remains." Feb. 17th, no pain along the spine, and bed-sores healing. "March 8th, improving slowly in both motion and sensation. Able to retain urine a short time and the incontinence of fæces has ceased. Bed-sores almost healed." March 15th, "still in bed, but now perfect control over legs. Still hyperæsthesia of feet. Incontinence of urine unaltered, but bowels are regular. Sores healed. No catamenia since accident, though regular before." "The condition having appeared," remarks Mr Hutchinson, "two days after the injury, led him to believe that there was some inflammation of the grey matter at the lower part of the spinal cord, set up by the concussion of the fall. The condition of inherited syphilis, however, introduced a new element; but although paraplegia was not uncommon in acquired syphilis, it was not common in the inherited form, and the history of the injury seemed to point to that being the cause of the disease. The treatment consisted at first in a slight mercurial course, afterwards in the administration of small doses of iodide of potassium, with of course rest in bed."

We would not lay greater stress than does Mr Hutchinson himself on the inherited syphilis of this patient, nor on the fact that recovery took place under the administration of anti-syphilitic drugs; but these facts to some extent detract from the value of the case, and make it less unequivocal than it would otherwise have been. A converse explanation is at hand, that had it not been for the syphilitic element the concussion lesion of the cord would not have run on to myelitis, and would have been therefore unrevealed. At the same time the original lesion may have been a small hæmorrhage upon the cauda equina and the cord, causing "pins and needles," insufficient of itself to induce immediate paraplegia; and the inherited vice of syphilis only paved the way for the supervention of inflammatory action from an otherwise inadequate cause. The injury at any rate seems to have been quite at the lower part of the back, and the question here arises which has been mooted before, is the lower dorsal and lumbar region of the spinal cord, surrounded by the nerves of the cauda equina and filling more closely than elsewhere the vertebral canal, more liable itself to suffer, independently of external hæmorrhage, from blows upon this part of the back? Recorded cases are too few to admit of any decided opinion, but the point is well worthy of consideration and should be borne in mind in an inquiry into "concussion of the spinal cord." A case, somewhat of the same kind, has been recently recorded by Dr Sharkey ('Brain,' April, 1884, p. 99), under the heading "A Fatal Case of Concussion of the Spinal Cord," where a girl met with a severe blow on the back-bone, and was admitted into hospital with retention of urine and complaining of extreme pain across the sacrum and of numbness over the left buttock. At the end of a week incontinence of urine and fæces set in and continued till death, and there were symmetrical bed-sores on the buttocks. She died exhausted in about six weeks, the bed-sores having during the last weeks of life begun to heal. No pathological changes were found in the spinal cord, other than those which were obviously of long standing, and which were associated with congenital abnormality of the brain and limbs, a circumstance which in our judgment seems to lessen the value of this case as an undoubted example of concussion lesion. Dr Sharkey, however, believed it to have been so, and we should hesitate to differ from the opinion of so able a pathologist. There is nothing, however, to be found in the record of his case at variance with the views which we have deduced from an examination of the many published cases of concussion of the cord. He thinks that the focus of the disease may have been very limited in extent and may have therefore escaped microscopical investigation, and that had there been an opportunity of examining the cord shortly after the accident the lesion might have been more easily found. The lesion "would probably have ended in the recovery of the patient, had not secondary disease of bladder and kidneys unfortunately proved fatal."

In an earlier part of this chapter we have quoted cases from different authors which, if they can be considered analogous to any cases of concussion of the brain, should be placed in this division of concussion of the spinal cord; and in quoting each of them we have named the objections, in no case we hope unreasonable, which have appeared to us to weigh against regarding them as cases of simple "concussion lesions of the spinal cord." That some of them, notably those by Boyer, are not easily explicable we do not for a moment deny, but how meagre are the histories given, how many important points are omitted, and how inferior, may we not add, in all probability were the methods of *post-mortem* examination to those alone deemed adequate in the present day! Nothing is easier than to accept such cases without examination or remark, and to note them down at once as examples of "concussion of the spinal cord," but we must not rest satisfied with this mode of collecting evidence, nor would it, indeed, suffice in an inquiry so important as this branch of our subject demands. Literature and clinical experience seem, indeed, to afford us little ground for thinking that even in this extensive class (Class B, p. 29) the analogy holds generally good between "concussion of the brain" and "concussion of the spinal cord."

In connection with this division of our subject the following case is of more than passing interest. It is entitled, "A case of concussion-lesion with extensive secondary degenerations of the spinal cord, followed by general muscular atrophy," and was brought by Dr

Bastian before the Royal Medical and Chirurgical Society on June 25th, 1867, in whose 'Transactions,' vol. 1, it is to be found.

"Jeremiah C—, aged 26, admitted into the accident ward of St Mary's Hospital on July 7th, 1866; about a week ago was sleeping on the top of an unfinished hay-rick, twenty-five feet in height, and whilst asleep rolled off, falling on his back. He found himself at once unable to move, and was conveyed to the Barnet Union, where he remained till, at his own request, he was removed to this hospital." It is unnecessary to give the exact words in which his various symptoms are recorded; let it suffice that there were imperfect paraplegia, paralysis of the bladder, partial palsy of the right arm, purely diaphragmatic breathing, and large bed-sores. He complained of soreness and stiffness in the neck, and of slight pain in the neighbourhood of the first and second dorsal vertebræ, but no fracture or displacement could be detected there. Tenderness at this part was really very slight, for considerable pressure and percussion were scarcely complained of. So he lingered, with but trifling variation in the symptoms, though during the last three months of his life he wasted perceptibly day by day, in spite of a nourishing diet with plenty of stimulants. He died on December 31st, 1866. The body was emaciated to a most extreme degree. There was no displacement or irregularity in any part of the spinal column, and the spinal cord was in no way compressed. No naked-eye changes whatever were to be discovered in the spinal cord. "The fact that no naked-eye appearances of disease could be detected in the cord at the post-mortem examination, seeing that extensive deviations from the normal structure were subsequently found to exist, is a subject of much interest in connection with the numerous instances in which pathological changes have been looked for in this organ and have not been recognised. . . . With the absence of the ordinary naked-eye characters of disease, with no deviation from the normal consistence, colour, or symmetry of the organ, it is not so much a subject of wonder that pathological changes, complying with these conditions, should escape detection even after a careful examination." After a *résumé* of the knowledge then existing of the origin of secondary degeneration of the spinal cord, and after giving an account of his own method of preparing the spinal cord for examination, Dr Bastian records in minute detail the microscopical appearances seen in this case. Here we would more especially direct attention to the grosser, more immediate, and primary lesions caused by the fall of twenty-five feet. "A transverse section of the hardened cord, through the upper part of the cervical enlargement (apparently corresponding with the interval between the fifth and sixth cervical nerves), showed a large rupture extending obliquely from before backwards across the grey matter of the right side." Other sections revealed "the sheaths of blood-vessels filled with amorphous granules of blood-pigment, of a dark olive-yellow colour, whose presence clearly indicated an original rupture of blood-vessels in this situation." Sections of the cord showed other and independent ruptures, and Dr Bastian writes: "It will be seen that the principal one of the original lesions or ruptures of the cord was situated in the upper part of the cervical enlargement, though there is every reason to believe that one or two other important lesions must have been situated in the portion of the cord immediately above this, which was unfortunately not preserved." Further, we read, "from the fact that in different parts of its circumference, in different sections, I have seen blood-vessels, or rather the sheaths of blood-vessels, perfectly loaded with altered blood-pigment, it seems most probable that several of the small blood-vessels supplying this portion of the grey matter had been ruptured by the original concussion, leading to effusions of blood into their sheaths, and hence obliteration of the vessels themselves from external pressure. The vascular supply to this portion of nerve-tissue being cut off or seriously diminished the tissue underwent a process of softening, which, at the period of the man's death, showed itself in the stage of repair."

That after such lesions as are here described, and as were seen or were suspected to exist in other parts of the cord, there should have been secondary degeneration, is not to be wondered at when we consider the exceeding delicacy of structure of the spinal cord.* These degenerations, consisting in the main of atrophied nerve-fibres, new connective tissue elements and granulation corpuscles, are fully investigated by Dr Bastian, and their physiological and pathological features pointed out, and we must refer to the paper itself for his exposition thereof. Here, however, we are more especially concerned with the fact that lesions—definite and discoverable after death—were caused by the original fall. The case presents innumerable points of interest. " In the first place, because from a concussion not more severe than might occasionally be experienced in a railway accident, the most unmistakeable and even extensive lesions of the spinal cord were produced at the time, and recognised after the patient's death. . . . Much interest also attaches to the distribution of the areas of secondary degeneration, on account of the bearing which this has upon the physiological anatomy of the spinal cord; and also to the histological nature of the changes produced, since these serve not a little to elucidate the real nature of cerebral or spinal *ramollissement*. And lastly, the gradual supervention of a general muscular atrophy in conjunction with the wasting of a portion at least of the great sympathetic system lends an additional interest to the consideration of this important case." In connection with

* And its proneness, moreover, to undergo degeneration *systematically*.

this last and most interesting observation it may be added, " that a careful inspection of the great semilunar ganglia and a comparison of them with others removed from patients dying of different diseases," enabled the author to say that they were " undoubtedly atrophied," contained " a larger proportion of thin fluid fat" than is usually met with, and that too in a body remarkable for the almost total absence of fat in it; and that the "ganglion cells seemed to contain rather more than their usual amount of pigment."

Whether the peculiar circumstances of the fall, whereby the whole spinal cord was caught at a disadvantage unusual in commoner modes of accident, conduced to the cord lesions in this particular case, it is possible only to surmise. The case is one of unusual interest and rarity; and it is not unworthy of observation that though Dr Bastian in his opening remarks somewhat invited the publication of similar cases by speaking of this one as a concussion lesion, such as might be met with after railway collisions, there has been as yet no record of any case at all comparable with it. It probably remains unique, unless, indeed, the extraordinary cases of Boyer were of the same nature, and the lesions were likewise undiscoverable without microscopic aid.

We come lastly to the third class of cases of concussion, viz. (C. vide *ante*, p. 29) those "*rapidly fatal* with concussion symptoms, where brain-lesions are almost invariably found after death, with or without damage to the membranes and the skull." Here we trench upon lesions and injuries of the greatest gravity, and it becomes almost impossible to compare cases of brain injury of this class with cases "rapidly fatal" after injury to the spinal cord. Examples thereof have been mentioned in the last section of this subject, and we there remarked that the time of death will vary according to the part of the cord which has been damaged. Unless we ought here to include the rapidly fatal cases of Boyer, we know of none which can be placed in this division, of which it may not be said that death has been caused by irrecoverable shock; or in which some very serious structural damage has not been occasioned to the cord by pressure of the displaced vertebral column; or in which, as in the cases of Mayo and Gull (vide *ante*, pp. 43 and 44), there has not been rupture of some vessel, the bleeding from which has rapidly annihilated the vital functions of the spinal cord.

We have thus been able to bring together a considerable number of cases of so-called concussion injury of the spinal cord, and amongst them we have seen how few there are in which there has not been damage likewise to the spinal column. In many of the cases we have given reasons for doubting the propriety of regarding the lesion in the cord as due to "concussion" *per se*, and we have shown in how few is the evidence of the lesion being due to concussion alone as free from question as are the ordinary concussion lesions of the brain. We have endeavoured upon anatomical grounds to show that the spinal cord shares but little of the risk of the brain to suffer lesion from blows directly inflicted upon its bony covering; and we have appealed to the unwritten experience of surgeons as to the rarity of lesions of the spinal cord in the absence of injury to the form, structure, and integrity of the spinal column. Although in many of these cases the physical signs of injury to the column are absent during life, yet, nevertheless, the lesion in the spinal cord is so commonly fatal that opportunity for arriving at an exact knowledge of the structural damage is usually not long delayed. We have further sought to inquire into the supposed analogy between concussion of the brain and concussion of the spinal cord, and have found that even if the analogy does in any case hold strictly good, it is only in the very rarest instances that it can unequivocally be maintained. And this fact is very prominent, that there is no evidence to show that the spinal cord can receive concussion injury without the manifestation of undoubted symptoms, or that the cord itself can meet with structural traumatic lesion without the appearance of those symptoms immediate upon the injury.

The evidence thus collected in a wide range of literature is singularly important at the outset of our inquiry : for if the spinal cord be, as we believe with all justice and fairness it may be said to be, so free from risk of concussion injury owing to its unrivalled security in the spinal canal, it seems highly improbable that it should be especially liable to suffer injury in any single kind of accident such as railway collisions, no matter how trivial they may be, and even though no damage has been inflicted on or near the spinal column. The improbability seems great, but far different might be the reality ; and we must, therefore, in the next place, direct our attention to the evidence which is to be gained as to the liability of the cord and its coverings to suffer injury in the collisions which are frequent forms of accident in the present day, and the results of which become so often the subjects of medico-legal inquiry.

CHAPTER II

"CONCUSSION OF THE SPINE"

I

HAVING dealt thus far with the general surgical aspect of concussion lesions of the spinal cord, it behoves us to treat, in the next place, of those spinal injuries which are, or are said to be, induced by the shock, concussion, or jars of railway accidents. In examining the question of " concussion of the spinal cord," we have hitherto omitted to refer in any detail to the opinions of more recent surgical authors, for it appears to us that their views have been influenced, and in some cases directed, less by any extended experience of their own, than by the writings of Mr Erichsen, who published in 1866, 'Six Lectures on certain obscure Injuries of the Nervous System, commonly met with as the result of Shocks to the Body received in Collisions on Railways ;'* and who, in a later and better-known work on 'Concussion of the Spine, Nervous Shock, and other obscure Injuries of the Nervous System in their Clinical and Medico-legal aspects,'† has laid before the profession, of which he is so distinguished a member, the results of his "more recent and extended experience." In this later work the "six original lectures have been incorporated, not, however, without much alteration, and eight new Lectures have been added." It is from these two works, containing his earlier and his later utterances, that

---

* Longmans and Co., 1866.
† Ibid., 1875, 2nd edition, 1882.

much, we believe, which has been written in recent medical and surgical works upon spinal concussion injuries, has been drawn, and we must, therefore, carefully investigate the doctrines which Mr Erichsen lays down in his later work on ' Concussion of the Spine.'

A difficulty meets us, even on the outside of the book, in the major title which has been chosen for it. There is something so altogether indefinite in the expression " concussion of the spine "—the " spine " being so commonly used as a comprehensive term for muscles, ligaments, bones, joints, membranes, spinal fluid, spinal marrow and nerves going off from it—that when we speak of "concussion of the spine" we must perforce use an expression scientifically inaccurate, and either more or less than adequate to describe a lesion affecting one only of the elements of which the "spine" is composed. "'Concussion of the spine' often used is objectionable as a title. We do not speak of concussion of the skull."* An " expression that is in itself perfectly definite, and that admits of no ambiguity in the mind of a medical man, may present a very different meaning to one who does not possess the requisite amount of anatomical or pathological knowledge to be able correctly to appreciate its true purport. Thus, for instance, the word 'spine' is used by an anatomist as signifying only the column, whereas a non-medical man will usually employ it as including the cord as well as its enclosing case."† These are the author's own words; and it is not an inapposite question to ask, whether " concussion of the spine " really means concussion of the vertebral column only, or does it signify concussion of any one or more, or all of the structures which compose or lie within it?

This is no merely verbal objection to the phrase. When we reflect how prone are injuries to the back, or to the "spine of the back," as it is not uncommonly called, to

* 'Holmes' System of Surgery,' 2nd. Ed., vol. ii, p. 370, footnote.
† 'Concussion of the Spine,' p. 321.

become the subject of medico-legal inquiry, the importance is doubly obvious of using names and titles which shall accurately express the nature of the injury and the part which has been injured. It is idle to deny that the temptation to exaggeration and imposture gives a tone and colour to a very large proportion of the cases of injury which involve litigation, or which are entitled to pecuniary compensation. "An extensive experience in railway compensation cases," writes Mr Erichsen (op. cit., p. 287), " will probably impress you more with the ingenuity than with the honesty of mankind. A history of deception practised on railway companies by alleged sufferers from accidents upon their lines would form a dark spot on the morality of the present generation." And for anyone to employ for the common injuries received in railway accidents a title which may now mean this, and now that, and very often may mean nothing at all, is to run a risk, it seems to us, of either playing into the hands of those who are using dishonest means to enhance their claims, or of seriously misleading those who, from lack of experience and opportunity, are ignorant of the symptoms and of the pathology of diseases of the spinal cord. But there the title is, and it behoves us to inquire what the author means by it.

"If the brain,"* writes Mr Erichsen, " is liable to suffer serious primary lesion and protracted secondary disease from the infliction of slight and, perhaps at the time, apparently trivial injuries to the head, the spinal cord is at least *equally prone to become functionally disturbed and organically diseased from injuries sustained by the vertebral column.*" His object, therefore, is to direct " attention to certain *injuries of the spine* that may arise from accidents, that are often apparently slight, *from shocks to the body generally*, as well as from blows inflicted directly upon the back ; and to describe the *train of progressive symptoms that lead*

* Op. cit., p. 1, *et seq.* Italics our own.

up to the obscure, protracted, and often dangerous diseases of the spinal cord and its membranes, that sooner or later are liable to supervene thereon. . . . These injuries of the spine and of the spinal cord occur not unfrequently in the ordinary accidents of civil life—in falls, blows, horse and carriage accidents, injuries in gymnasiums, &c., but in none *more frequently or with greater severity than in those which are sustained by persons who have been subjected to the violent shock of a railway collision.*"

The author has no " wish to make a distinction in injuries of the spine according to their causes, and still less to establish anything like a speciality of 'railway surgery,'" but he speaks of injuries of the spine from railway collision more especially in his lectures, " because injuries of the nervous system of the kind we are about to discuss have become of much practical importance from the *great frequency of their occurrence,** consequent on the extension of railway traffic, and because they are so frequently the cause of litigation." The† " more serious injuries to the nervous system, whether affecting the brain, spinal cord, or peripheral nerves—whether arising from wounds, from fracture of the skull, or fracture and dislocation of the spine—have been so thoroughly studied by all practical surgeons that little now remains to be said" upon them, and with them he has " at present no concern." " But the primary effects and the *secondary results of slight* injuries to the nervous system* do not appear, as yet, to have received that amount of study and attention on the part of surgeons that their frequency and their importance alike demand. The neglect with which these cases have hitherto been treated appears the more extraordinary when we consider the peculiar interest that their phenomena always present, and the important position that they have, of late years, assumed in medico-legal practice." While, then, the purport of Mr Erichsen's lectures is partly to supply " a

* Italics our own. † Op. cit., p. 3.

missing chapter in medical jurisprudence," and partly "with the view and in the hope of clearing up some of the more obscure points connected with these injuries," one of his main objects is to show that shocks to the nervous system arising from railway accidents do not stand in a different category from injuries induced by other causes in civil life, and that "precisely the same effects may result from other and more ordinary injuries." If there be a difference it is to be found in the alarming nature of railway collisions generally. "The[*] cause is special, and the results are peculiar; but though peculiar they are not so unlike those arising from other accidents as to justify us in regarding them as being in any essential respect distinct and different. The peculiarity of these obscure injuries of the nervous system caused by railway shocks is sufficiently great, however, to warrant us in grouping them together, and considering them as a whole in a separate chapter in the great book of surgery." [†]"But although," he proceeds, "the intense shock to the system that results from these accidents naturally and necessarily gives to them a terrible interest and importance, do not for a moment suppose that these injuries are peculiar to or solely occasioned by accidents that occur on railways. There never was a greater error. . . . It is an error begot in egotism and nurtured by indolence and self-complacency. It is easy for a man to say that such and such a thing cannot exist, because 'I, in my large experience at our hospital, never saw it,' whereas, if he would take the trouble, he would find, by the study of their works, that surgeons of equally large, or perhaps of far greater, experience in their generation have seen and described it. Formerly this opinion might have been excusable; it is no longer so. The comparative rarity of these obscure injuries of the nervous system in ordinary hospital practice and in private caused them either to be entirely over-

[*] Op. cit., p. 5.   [†] Op. cit., p. 6.

looked, or to be regarded as mere surgical curiosities," but now-a-days, occurring as they do " too frequently in groups, sometimes of scores at a time, they have been brought under the observation of every surgeon, and their symptoms, prognosis, diagnosis, and treatment form an important part of the professional occupation of practitioners in every part of the country." "Surgical literature of the past century" shows that "cases of slight accidents to the spine or head, followed by serious persistent or fatal results, were not unknown;" and "in the writings of Sir A. Cooper himself, in those of his predecessors and contemporaries, especially of Boyer, of Sir C. Bell, and at a later period of Ollivier and Abercrombie," are to be found "isolated cases" which "prove incontestably that precisely the same series of phenomena that of late years have led to the absurd appellation of the 'railway spine' had followed accidents, and had been described by surgeons of the first rank in this country and in France, a quarter of a century and more before the first railway was opened, and that they were then generally recognised as arising from the common accidents of civil life. The only difference is that accidents have greatly increased in frequency and intensity since the introduction of railways, and *these injuries have become proportionally more numerous and more severe.*"* We have purposely italicised these words, for, as we have followed Mr Erichsen —and nothing of essential moment has been omitted from our quotations—it seems to us that the words " *these injuries* " refer in the passage last quoted to a class of cases differing somewhat from those which were indicated when he was earlier speaking of " slight† injuries to the *nervous* system*" and of‡ " 'shocks' to the *nervous* system* arising from railway accidents," all of which, however, he considers should not " stand in a different category from accidents occurring from other causes in civil life," because " precisely

* Italics not in original. † Op. cit., p. 3. ‡ P. 5.

the same effects may result from other and more ordinary injuries."

Let us be clear upon this point, for we shall learn, as we proceed, how vast and how really important is the distinction between these sets of cases. That "*these injuries*" now comprise "slight accidents to the *spine or head*, followed by serious persistent or fatal results," seems no less than certain when Mr Erichsen goes on to tell us that we may go further back than the writings of these great men whom he has named, and find scattered here and there throughout medical literature some most interesting cases that bear upon this very point.

And here it is necessary to quote a long passage. "If* you take up the third volume of the 'Medical Observations and Inquiries' you will find that in 1766, more than one hundred years ago, a case is related by Dr Maty of 'a palsy occasioned by a fall attended with uncommon symptoms,' which is of so interesting a nature, and which bears so closely upon our subject, that I feel that I need offer no apology for giving you an abstract of it here, although as it occurred between sixty and seventy years before the first railway was opened in this country it might at first appear to have less relation to railway accidents than it really has, for in its course and symptoms it is identical with many of them. This case, which is given at length, and which I shall abstract from the original, is briefly as follows:

"Count de Lordat, a French officer of great rank and much merit, whilst on his way to join his regiment in April, 1761, had the misfortune to be overturned in his carriage from a pretty high and steep bank. His head pitched against the top of the coach; his neck was twisted from left to right; his left shoulder, arm, and hand were much bruised. As he felt at the time little inconvenience from his fall he was able to walk to the

* Op. cit., p. 10, *et seq.*

next town, which was at a considerable distance. Thence he pursued his journey, and it was not till the sixth day that he was let blood on account of the injury to the shoulder and hand. The Count went through the fatigues of the campaign, which was a very trying one. Towards the beginning of the winter (at least six months after the accident) he began to find an impediment to the utterance of certain words, and his left arm appeared to be weaker. He underwent some treatment, but without much advantage; made a second campaign, at the end of which he found the difficulty in speaking and in moving his left arm considerably increased. He was now obliged to leave the army and return to Paris, the palsy of the left arm increasing more and more. Many remedies were employed without effect. Involuntary convulsive movements took place all over the body. The left arm withered more and more, and the Count could hardly utter a few words. This was in December, 1763, two years and a half after the accident. . . . In October, 1764, three years and a half after the fall, Dr Maty saw him. 'A more melancholy object,' he says, 'I never beheld. The patient, naturally a handsome, middle-sized, sanguine man, of a cheerful disposition and an active mind, appeared much emaciated, stooping, and dejected. He walked with a cane, but with much difficulty, and in a tottering manner.' His left arm and hand were wasted and paralysed; his right was somewhat benumbed, and he could scarcely lift it up to his head. His saliva dribbled away; he could only utter monosyllables, 'and these came out, after much struggling, in a violent expiration, and with a low tone and indistinct articulation.' Digestion was weak; urine natural. His senses and the power of his mind were unimpaired. He occupied himself much in reading and writing on abstruse subjects. No local tumour or disease was discoverable in the neck or anywhere else. From this time his health gradually declined, and he finally died on the 5th March, 1765, nearly four years after the accident.

"On examination after death the pia mater of the brain was found 'full of blood and lymph;' and towards the falx there were some marks of suppuration. The medulla oblongata is stated to have been greatly enlarged, being about one third larger than the natural size. The membranes of the cord were greatly thickened, and were tough. The cervical portion of the cord was hardened, so as to resist the pressure of the fingers. 'From these appearances,' says Dr Maty, 'we were at no loss to fix the cause of the general palsy in the alterations of the medulla spinalis and the medulla oblongata.' The twisting of the neck in the fall had caused the membranes of the cord to be excessively stretched and irritated; the morbid changes then extended by degrees to the spinal marrow, which, being thereby compressed, brought on the paralytic symptoms."

"This case," continues Mr Erichsen, "is of the utmost interest and importance, and though it occurred and was published more than a century back, it presents in so marked a manner the ordinary features of 'concussion of the spine' that it may almost be considered a typical case of one of those accidents." After naming the several points which he deems of interest in the case, Mr Erichsen concludes:*—"You will find, as we proceed in the investigation of this subject, that the symptoms, their gradual development, and the after-death appearances presented by this case, are typical of the whole class of injuries of the spine grouped together under the one common term 'concussion' from whatever cause arising." A terrible case, forsooth, to be pointed out as "typical;" and typical of what? "Of the *whole class*† of injuries of the spine grouped together under the one common term 'concussion' from whatever cause arising;" typical, in fact, of the cases

* Op. cit., p. 13. † Italics are our own.

once so rare that surgeons like Astley Cooper and others of the largest experience did not "appear to have seen a sufficiently large number to treat specially of them," but now, alas! "proportionately more numerous and more severe," because "accidents have greatly increased in frequency and intensity since the introduction of railways."

Is it possible that such a case as this can be rightly termed a "slight injury to the nervous system," or a "shock to the nervous system," when we note how the accident occurred? And yet the case is "typical of the whole class of injuries of the spine grouped together under the one common term 'concussion' from whatever cause arising." It is typical therefore of some slight injury to one of the ligamentous structures only of the vertebral column; for in defence of the term "concussion of the spine," Mr Erichsen writes in his last edition (op. cit., p. 13), "In concussion of the spine we have not only, and not even necessarily, an injury of the cord, but also, and perhaps solely, an injury of the osseous, fibrous, ligamentous, and muscular structures that enter so largely into the conformation and support of the vertebral column—of the nerves that pass across it—and of the membranes included within it." Small wonder when a man gets a slight sprain of his vertebral column in the most trifling collision on a railway that, labouring under the belief he has received a "concussion of the spine," his anxiety should be needlessly great and prolonged if he learns that the result of the injury in this oft-quoted case of the Count de Lordat is the typical result of such a "concussion of the spine" as he has himself received, even though he does not increase his risks of protracted illness by going through two campaigns.

The dust of the unhappy Count must have undergone a "molecular disturbance" in its tomb when, in the very opening pages of the book, this painful history was transcribed as "typical of the whole class of injuries" grouped under the term "concussion of the spine."

Sir Charles Bell alludes to this very case of the Count de Lordat, which is recalled to his mind, he writes, by one he had himself recorded under the heading "Injury of the Spinal Marrow from a Hurt on the Spine,"* where a man fell from some steps in hanging a curtain, and struck the lower part of his spine against the corner of a table. "The bruise was severe, but he got the better of it by the usual remedies, and in the usual time. It was some months after that he began to feel a want of power over the lower extremities," which the man, however, did "not attribute to his former accident, the more especially as so long a time had elapsed before these symptoms appeared." Bell also refers to another case where a man had "uneasiness† and defect of action in the lower extremities" after falling "forty feet down a shaft;" and of all three cases, he remarks: "Upon examination" (*i.e.* of the Count de Lordat) "the membranes of the spinal marrow were found thick and tough, and the marrow itself had acquired an extraordinary degree of solidity. The symptoms of the two slighter cases of palsy . . . . are, I imagine, to be explained on the same principle, viz. the injury to the soft envelope of the spinal marrow, and the accession of inflammatory thickening."‡ Assuming that in these cases the injury was the real cause of the symptoms, the explanation of Bell is doubtless correct; but there must always be a difficulty in tracing to an accident those symptoms or diseases which have shown themselves only after many months, and with an intervening period of perfect health. *Traumatic* inflammation of the spinal membranes is, be it remembered, an exceedingly dangerous affection, and one prone to run a very rapid course, for, as Bell remarks, "the membranes of the spinal marrow are

* 'Institutes of Surgery,' vol. i, 1838, pp. 154-5.
† Op. cit., p. 153.   ‡ Op. cit., p. 156

the most susceptible of inflammation and suppuration of the whole frame; not exceeded by those of the brain itself, of which they are prolongations;"* and when we now-a-days see cases of chronic meningitis such as these, we are especially careful to inquire into the history of syphilis, a disease now known, but at that time not known, to be a very common cause of this particular form of meningeal inflammation and thickening. Such a consideration must be borne in mind in a study of these cases—even in that of the Count de Lordat—and in the absence of special record or knowledge on this point, they fail to provide us with unequivocal examples of spinal meningitis, with its attendant consequences, *caused* solely by the injuries which each received. It is, however, well recognised that injury —even when apparently slight—of the vertebral column may be the real cause of a slow meningeal thickening or inflammation which may involve—and reveal its presence by involving—the nerves coming off from the spinal cord, or ultimately even the spinal cord itself. But uncomplicated cases of this kind are exceedingly rare, as indeed is shown most strikingly by the absence of so untoward a result in the numerous cases of injury to the back, and of sprain of the vertebral column, received in railway collisions.

Mr Erichsen, in the next place, goes on to deal more specially with the effects, immediate and remote, of† "those forms of concussion of the spinal cord which follow a *severe* degree of external violence applied to the vertebral column," and although admitting that it is by no means easy to give a clear and comprehensive definition of the term " *concussion of the spine,*" he proceeds to explain that it is a phrase "generally adopted by surgeons to indicate *a certain state of the spinal cord* occasioned by external violence; a state that is independent of, and usually, but

* Op. cit., p. 157.
† Op. cit., p. 15, *et seq.* Italics not in original.

not necessarily, uncomplicated by, any obvious lesion of the vertebral column, such as its fracture or dislocation, —a condition that is supposed to depend upon a shake or jar* received by the cord, in consequence of which its intimate organic structure may be more or less deranged, and by which its functions are certainly greatly disturbed, so that various symptoms indicative of loss or modification of innervation are immediately or remotely induced." "Concussion of the spine" is a term he says, in which surgeons and writers on nervous diseases† appear to include various distinct pathological conditions, having "only this in common, that they are not dependent upon an obvious external injury of the spine, such as the laceration or compression of the cord by the fracture or dislocation of a vertebra."‡ To cases recorded by these various authors he briefly refers. Some of them we have ourselves already examined, and have expressed our objections to receiving them as cases of "concussion of the spinal marrow," a term, by the way, which these writers use in notable preference to "concussion of the spine." "Sir A. Cooper," writes Mr Erichsen, "relates two cases of concussion of the spine, one terminating at the end of ten weeks in complete, the other in incomplete, recovery."§ Sir Astley in reality records them under the heading "Concussion of the Spinal Marrrow,"‖ and if we turn to them we find that the objections made to other cases apply with no less force to them. The first case is that of a man who received a "severe blow from a piece of wood which

---

* The kind of jar, for example, which the author has previously (p. 6) named. "Perhaps the one circumstance which more than any other gives a peculiar character to a railway accident is the thrill or jar, the *ébranlement* of French writers, the sharp vibration, in fact, that is transmitted through everything subjected to it."

† Cooper, Mayo, Bell, Boyer, Abercrombie, Ollivier, *e.g.*

‡ Op. cit., p. 16.

§ Op. cit., p. 17.

‖ 'Dislocations and Fractures of Joints,' 8vo ed., pp. 526, *et seq.*

fell upon his loins and knocked him down;" caused a "severe contusion and much deep-seated tenderness" at the site of the blow, and was followed by almost total paraplegia, from which in ten weeks he completely recovered. The second case is that of a "gentleman, who, by a fall from his gig, had received a severe blow upon his loins, and who had, at first, great difficulty in discharging both his urine and fæces, but he was relieved by fomentation and cupping." With all deference to the opinion of so illustrious an authority, Sir Astley Cooper, would, it seems to us, have more accurately placed his first case in Section III, "Injuries of the Spine, Extravasation into the Spinal Canal;" and in the second case we may fairly doubt whether there was any true paralysis or lost function of the cord at all, and whether the "difficulty," by no means an uncommon one, as we shall point out in the next chapter, after severe bruise and sprain of the lumbar muscles and ligaments, was not essentially due to the severe muscular bruising which the parts received.

Mr Erichsen's second lecture on the 'Effects of direct and severe blows on the Spine' contains a series of thirteen cases of injury to the spine which have been met with by himself in his own practice. They are most of them instructive examples of spinal injury occurring in the ordinary accidents of every-day life, from such accidents as a man may meet with in riding and driving, from felling trees, or from severe blows upon the spinal column in the course of laborious work, or from being knocked down in the street. It is, however, worthy of remark that of the thirteen cases recorded in this lecture, and of three more in Lecture III, 'On the Symptoms of severe Concussion of the Spine from Direct Violence,' only one[*] is a case of

[*] For Case 14 in Lecture III, 'On the symptoms of severe Concussion of the Spine from Direct Violence,' where a man received many and terrible injuries in getting out of a train before it had stopped, and falling between the platform and the carriage, and died on the

injury met with in a railway collision. A very singular disproportion, when we recall with what frequency these railway injuries are said by the author to occur, and when the object of the work is essentially to deal with and throw light upon the different kinds of injuries to which railway collisions give rise. For these very valuable cases we must refer our readers to the book itself; but there are three or four of them to which we would more especially direct attention.

To begin with " Case 2.* *Fall on back. Partial paraplegia. Cerebro - meningeal symptoms. Incomplete recovery.*"—A painter, æt. 30, fell with his ladder to the ground, a height of about thirty feet, and struck his back upon a gravel walk. His head was uninjured, but on admission to the hospital, in June, 1865, he was somewhat collapsed and cold. " There was no evidence of fracture either of spine or pelvis, but the back was ecchymosed to some extent about the centre of the dorsal region." His symptoms were those of partial paraplegia, both of motion and sensation, and he had perfect control over his sphincters, the urine being acid. He somewhat improved in August, when he was able to sit up, and in September, when he left the hospital, he was " emaciated, cachectic-looking, and could barely manage to walk and drag his leg, by holding on to the furniture or by pushing a chair before him." He slowly improved, but when seen ten months after the accident, †" he described himself as being languid, depressed, and as if going out of his mind. His memory had become very bad—at times all seemed a blank to him. When he went on an errand he often could not recollect what it was

---

fourth day, having amongst other lesions a dislocation of the cervical spines between the second and third vertebræ, is, of course, not an ordinary "railway accident" case at all, and may be excluded from the category of true collision injuries, in which there is the peculiar "thrill or jar."

* Op. cit., p. 20, *et seq.* † Op. cit., p. 22.

about; was always obliged to write it down. His thoughts were confused; he often mixed up one thing with another. He was very nervous and easily frightened. He dreamt much, and was told that he talked and cried in his sleep. He said he was 'not the same man that he was,' and thought he never would be. He could not do ordinary work as before the accident—only 'odd jobs.' He could not walk more than a mile, and could not carry a pail of water without great exertion." He had moreover aching and throbbing in the back, tenderness in the spine, and on either side of it; pain in the back increased by movement; difficulty in stooping, unsteady gait; numbness and "pins and needles" in the right leg and foot; muscæ volitantes and coloured spectra; distress at loud sounds; very acute hearing; function of bladder natural. At the end of two and a half years he was little, if any, better.

That a man who fell thirty feet on his back, and either from intra-spinal hæmorrhage or by undiscovered fracture or dislocation of the vertebræ, received some serious injury to the spinal cord, should suffer from this train of symptoms is not to be wondered at; nor would further comment be needed were it not for the passage which has been quoted in full. We shall do well at once to point out that this quotation very fitly describes the kind of complaints which are so frequently heard after the receipt of railway injury, with or without damage to the back at all. The same class of symptoms, which we ourselves have heard from a man who broke his jaw, and from another who had a simple fracture of the fibula, may be noted in others of the cases. Thus we read of a man (Case 9) who met with a direct blow on the back by a fall on to a pile of rough stones, and who felt no ill effects until three months afterwards—a fact which in itself makes it extremely doubtful whether his illness was really due to the injury—that* "the first symptom he complained of was loss of sleep. He was

* Op. cit., p. 37.

unable to sleep more than three hours at a time. He then suffered from extreme mental depression, became hypochondriacal and suicidal. He was extremely nervous, so that he could not with comfort be left alone." And in Case 10 also, called "Direct blow on back by fall downstairs—slow development of symptoms of meningeal irritation and of paralysis," the case of a woman aged 45, who had been injured three and a quarter years before, and whose whole history, we think, looks much more like that of an hysterical woman at the menopause than one of real organic disease as the result of injury, we read as follows: "She* complained that her memory was impaired, that she forgot dates; she could not recollect where she placed things, and occasionally she used the wrong word or forgot a particular word she wished to employ when talking. She was apt to lose the thread of her sentence so as to have to begin it again. Her sleep was greatly disturbed by dreams of a terrifying character. There were constant noises in the head, slight deafness of the right ear. Any sudden or loud noise, such as the crying of children or the falling of fire-irons, distressed her extremely." How often may these and like complaints be heard after railway injuries of any and every kind! Let us not forget that they are symptoms in no wise the direct result of disease or injury of the spinal cord, and that they are of no value whatever in determining the presence or the absence of organic disease of the important structures lying within the vertebral column.

Cases 12 and 13, which Mr Erichsen has previously remarked are "instances† of death following concussion of the spine," deserve our close consideration. Case 12 is the railway case and the only one in all probability of the series which became the subject of medico-legal inquiry, and it is entitled:‡ "Slowly developed spinal meningitis, from direct injury received in a railway collision, termi-

* Op. cit., pp. 38, 39.   † Op. cit., p. 29.   ‡ Op. cit., p. 40.

nating eventually in death." The accident happened to a post-office clerk on June 23rd, 1866, when he "was violently struck on the right side and loin against the edge of the table" in the post-office van. It is unnecessary to give in full the minute details of this case, suffice it that mixed up with symptoms which might perhaps be attributed to spinal meningitis, are many symptoms of emotional and functional disturbance to which at a later stage we shall more especially refer. The date is not given, but it would appear that his claim for compensation was arranged some time in the second half of 1867, at any rate, about a year after the accident, and the following paragraph closes the history:—"After* the conclusion of the legal proceedings connected with the case, I lost sight of the patient, who retired into the country; but Dr Waller Lewis informed me, in 1871, that he had eventually died from the effects of the accident. The particulars of the latter period of his illness and of his death could not be obtained." Eventual death is our common lot, and we cannot of course deny that this man died from the effects of the accident. We do deny, however, that there is sufficient evidence to support the statement. Every ailment of his future life, as we ourselves have seen in several instances, would certainly be attributed to the railway accident, and to that also his friends would doubtless attribute his death. Our astonishment, however, does not rest here, for if we go further with the author, and turn to page 305 of his book, we there learn that of the three modes in which "concussion of the spine may prove fatal" the third is, "after the lapse of several years by the slow and progressive development of structural changes in the cord and its membranes." And placed in brackets beside these words we find "(Case 12)." Is this the kind of evidence, in all seriousness we would ask, with which we ought to be satisfied, before we accept this, or any, case as an un-

* Op. cit., p. 43.

doubted instance of death resulting from injury received in a railway collision years before, or can we allow that it is a happy example of one of the modes in which " concussion of the spine" may prove fatal? It is the railway case, and there was no post-mortem examination. It is worthy of remark, moreover, that there were " legal proceedings " connected with it, presumptive evidence either that a totally different view of the case was taken by the medical advisers of the railway company, or that a claim was made out of all proportion to the *real injuries* received.

Lastly, we have " Case 13. *Severe contusion.** Paraplegia. Unsuspected laceration of intervertebral ligaments. Death on ninth day;*" not, it will be observed, immediately called "concussion of the spine," although at p. 304 Mr Erichsen writes: " It is certain that *concussion of the spine*† may prove fatal; first, at an early period by the severity of the direct injury (Case 13)." A man having been knocked down by a cab, "the horse falling partly upon him, and striking him on the neck with its knee," was admitted into hospital with complete paralysis of motion and sensation in the lower extremities, and in the trunk as high as the shoulders. "There was no inequality or irregularity about the spinous processes, or any evidence of fracture of the spine, but the patient complained of severe pain at the site of the bruise." Bed-sores formed, the urine became ammoniacal, and he died ten days after the accident. " On‡ examination after death the head and brain were found uninjured and healthy. On exposing the vertebral column, it was found that the sixth and seventh cervical vertebræ had been separated posteriorly. The vertebræ themselves, and their arches, were quite sound, but there was a fissure without any displacement, extending through the articulating processes on the left

* Op. cit., p. 43.
† Italics not in original.   ‡ Op. cit., p. 44.

side. A large quantity of blood was extravasated into the spinal canal, lying between the bones and the dura mater. There was a considerable quantity of reddish serous fluid in the arachnoid. The pia mater of the cord had some blood patches upon it on the lower cervical region. The cord itself was quite healthy." And then occurs this comment: "The fracture of an articulation without displacement was an accidental and insignificant complication, the real injury consisting in the extravasation of blood within the vertebral canal, which, by compressing the cord, induced the paralysis that ultimately proved fatal." Insignificant! The significance of the whole case seems to us to lie in the fact that there was a separation posteriorly between the vertebræ, which allowed of greater flexion in the easily-bending cervical spine, and that without this flexion there would in all probability have not been the large quantity of blood extravasated into the spinal canal, which pressed upon the cord. Significant, too, as teaching us what has indeed been alluded to before, that these separations, more especially in the cervical region, are very likely to be undiscoverable during life; and that when we meet with paraplegia occurring after *severe* injuries to the spine, and there be no direct evidence of damage thereto, there is yet strong presumptive evidence, from the lessons of the dead-house, that the vertebral column has itself been severely injured, and that from the immediate consequences of such injury the function of the spinal cord has been annulled and destroyed.

Mr Erichsen then goes on to deal in full with the "Symptoms of severe concussion of the spine from direct violence" (Lecture III); and we may fully agree with his remark as to there being* "every possible variety in the extent, degree, and relative amount of paralysis of motion and of sensation," which is met with after injuries of the cervical spine. "The condition," he further writes, "that

* Op. cit., p. 51. Italics our own.

is most frequently developed by a direct blow on the middle or lower dorsal, the lumbar or lumbo-sacral regions, is that of paraplegia. The symptoms presented by the patient who is thus paralysed below the seat of the *concussion of the spine* are necessarily those which result from such a disturbance, commotion, or lesion of the cord, as will occasion serious modification or complete suspension of its functions." When, however, we come to go through the various symptoms of the "primary and more immediate forms" of injury, arranged under the heads:* "1. Diminution or loss of motor power. 2. Rigidity and spasm of muscles. 3. Diminution or loss of sensation. 4. Perversion of sensation. 5. Loss of control over the sphincters. 6. Modification of the temperature of the limb;" we meet with some difficulty in always comprehending to what kind of injuries the author refers; and that more especially when he uses the expression "concussion of the spine.' Thus he writes:† "Pain and perverted sensations of all kinds are very common in cases of spinal concussion from direct blows on the back." And again, speaking of exalted sensation:‡ "There is a minor degree of this form of hyperæsthesia, consisting of the sensation of a cord tied tightly round the body, which is very common in severe blows, more especially in wrenches of the spine, and which seems to be dependent rather on pressure on the nerves by ligamentous strain than by bony fracture."

A case is then given as "a good illustration of this form of hyperæsthesia:" " Case 16. *Injury of the spine in lower dorsal region. Recovery with angular curvature.*"§ A man was seated on the top of a pleasure van, when the driver attempted to pass under a low archway. "The patient stooped forwards, the edge of the arch struck the lower part of his back, and he was thus dragged through the archway, forcibly doubled up and crushed between it and

* Op. cit., p. 52. ‡ Op. cit., p. 58.
† Op. cit., p. 57. § Op. cit., pp. 58, 59.

the top of the van. . . . . When seen, a few minutes after the accident, ... the left leg was paralysed, and there was intense hyperæsthesia over the lower part of the thorax and back, *i. e.* below the level of the third costal cartilage, as well as over the abdomen and upper two thirds of the thighs. . . . . . Opposite the spines of the tenth, eleventh, and twelfth dorsal vertebræ there was a considerable prominence, which terminated abruptly below in a depression." "On more careful examination," two days after the accident, "the limits of the hyperæsthesia were as follows. On the front of the left leg, it extended to three inches below the knee; on the right, to the upper border of the patella; posteriorly, it extended to the middle of the thigh in both limbs; above it began at the lower border of the ribs. . . . . He still complained of the sensation of having a band tied round the abdomen." He was discharged from hospital convalescent with an angular curvature of the spine, the ninth dorsal vertebra standing out most prominently. There can be no doubt that in this remarkable case there was very grave injury to the spinal column, and in all probability * " considerable contusion of the spinal cord," such contusion being directly induced, in our opinion, by the severe bend to which the column was subjected. The case fully confirms what has been said in the last chapter on the frequent connection between injuries of the cord and injuries of the vertebral column. In reference to this case it is right to notice that the term " concussion of the spine " is not immediately used by the author.

But further, in speaking of paralysis of the sphincters, Mr Erichsen says that this †" is an extremely uncertain symptom. It is sometimes met with in comparatively slight cases, especially when the blow leading to the concussion has been inflicted low down, in the lumbar and sacral regions. It is sometimes absent when both the lower limbs

* Op. cit., p. 61.   † Op. cit., p. 63.

are completely paralysed. If, however, the seat of concussion be about the middle dorsal vertebræ, and if the injury be severe, it is always present to a greater or less degree." Mark, in this passage the use of the word "concussion," and then let us observe it in the following quotation from the next page :—*" Priapism does not occur in concussion as it does after laceration and irritation of the cord." And again :—" In spinal concussion there is as a rule a fall of temperature. In laceration or crush of the spinal cord, consequent on fracture of vertebra, there is often a rise— the more so if the cervical spine is the seat of injury." It is, we think, obvious that there is a distinction involved in these passages which have been quoted, between cases of undoubted lesion of the spinal cord and cases of some other kind; and yet if we go on to a case where † " it is evident that not only the meninges of the cord, but the ligamenta subflava were torn through, and the arches of the vertebræ separated to such an extent that the softened and disorganised medulla found a ready exit through the gap thus made at the posterior part of the spinal column," we find this comment : "It is a point of much practical moment to observe that in this as in several other of the cases of so-called 'concussion of the spine,' there is in addition to the lesion of the cord, some serious injury inflicted on the ligamentous and bony structures that enter into the composition of the vertebral column, which, however, must be considered as an accidental complication, as it does not occasion, or even seriously aggravate, the mischief done to the medulla itself." And here again we cannot but express our wonderment that so serious a lesion of the spinal column as was present in this and other cases should be regarded as merely an accidental complication, neither occasioning nor seriously increasing the mischief done to the medulla itself. Ignore these injuries to the vertebral column, and in ignoring them assume that the spinal cord

* Op. cit., p. 64.   † Op. cit., p. 73.

has therefore been damaged by "concussion" as the brain may be by "concussion" *per se*, and the step is made easier towards the establishment of concussion lesions of the cord where the vertebral column has received no injury at all.

Not less remarkable than this last case—if we go back a few pages—is * " Case 5. *Direct blow on cervical spine. Paralysis of left arm ;*" spoken of a few lines before as a case which " shows that a† *concussion of the spine may be followed by paralysis of one limb only.*" A man, aged 62, had been struck, some months before Mr Erichsen saw him, by a branch which fell from a tree on the " left side of the neck, shoulder, and spine. There was no fracture or dislocation ; the severity of the blow was expended on the side of the neck, chest, and shoulder ; the head was not struck. The whole of the left arm instantly became paralysed, both as to sensation and motion, and had been so ever since." The muscles attached to the scapula and humerus were wasted, and also of the arm and forearm. "The limb was rigid, the joints could not be flexed without a very considerable amount of pain. The fingers were partly flexed, and sensation was entirely lost below the elbow. Above this part it was normal. He suffered severe pain along the course of the ulnar and median nerves, which came on in spasms, and was very intense. There was tenderness on pressure from the sixth cervical to the fifth or sixth dorsal vertebræ, and constant pain there." These were the symptoms, and this is the comment : " The case appeared to be one of paralysis of the nerves of the upper extremity, from a direct blow on the spine, about the region of the brachial plexus on the left side." Surely if there ever was a case of direct injury to the component cords of the brachial plexus this must have been one. The very fact that one limb alone was so extensively paralysed, both as to motion and sensation, is almost as conclusive

* Op. cit., pp. 29, 30.   † Italics not in original.

evidence as we could well wish for that injury to the spinal cord was not the cause of the palsy, and that the lesion must have been outside the spinal marrow, most probably of the nerves themselves. That the man had a blow on his spine as well as upon the brachial plexus is altogether beside the point, and accurate nomenclature, we can hardly help thinking, would have distinctly separated the effects of the blow upon one part from the effects of the blow upon another. What, however, has the brachial plexus to do with the spine? Let that pass, if only we may most strenuously object to such a case being called "concussion of the spine." Used now to indicate this injury, and now that, here signifying the cause, there the effect, by a writer so distinguished as Mr Erichsen, it is little wonder that a wider application even has been given to the term, and that, as we shall see by and by, "concussion of the spine" is used almost indiscriminately both in and outside the medical profession to indicate the injuries which are received in collisions and which become the subject of medico-legal inquiry, although the spinal column and its contents have met with no damage at all. It appears to us nothing less than lamentable that in laying before the profession and the world the results of his experience upon this subject, and writing from the high vantage-ground of an assured position both as a surgeon and as a teacher of surgery, Mr Erichsen should not have been more clear, more explicit, and less ambiguous in the use of the phrases which he has employed.

II

The consideration given thus far to "concussion of the spinal cord," and to the instances of severe injuries of the spine recorded by Mr Erichsen, is not more than the subject itself deserves, and is a very necessity if we are to examine, understand, and appraise at its right value the evidence which Mr Erichsen adduces as to " concussion of the spine from slight or indirect injury."

As has been abundantly pointed out in the preceding pages, there is but scanty proof of the liability of the spinal cord to suffer from *concussion* pure and simple in the absence of simultaneous injury to the spinal column, the exceptional cases being extremely rare.

And if uncomplicated " concussion " lesions be so exceptional, and the spinal cord be, as we believe the common experience of surgeons proves it to be, the most securely protected of all the organs of the body, it seems most improbable that it should be prone to incur lesions due solely to indirect and general concussion. Were this really so, it would indeed be most remarkable, and a fact of the greatest importance as far as the injuries received in railway collisions are concerned. It is of the utmost moment, therefore, that we should endeavour clearly to understand what Mr Erichsen has written upon this branch of the subject. He goes on in Lecture IV, " On concussion of the spine from slight or indirect injury," to consider a class of cases where the injury upon the back is * " *either very slight in degree, or in which the blow, if more severe, has fallen upon some other part of the body than the spine, and in which, consequently, its influence upon the cord has been of a less direct and often of a less instantaneous character.*" It is obvious that when such accidents have happened, the difficulty of prognosis—a point, be it remembered, of vast importance in medico-legal inquiries—must often be very great, and especially so because in †" consequence of the length of time that often intervenes between the occurrence of the accident and the production of the more serious symptoms, it becomes no easy matter to connect the two in the relation of cause and effect." We must pass over the examples, several of them cases of

* Op. cit., p. 77. Italics our own.   † Op. cit., p. 78.

railway injury, which Mr Erichsen gives in Lecture IV of "Concussion of the spine from slight or indirect injury," in Lecture V of "Concussion of the spine from general shock," and in Lecture VI of "Sprains, twists, or wrenches of the spine," and proceed to that which is far more important than the cases themselves, his teaching "On the mode of occurrence of shock," and "On the pathology of concussion of the spine."*

After referring to the "disproportion that exists between the apparently trifling injury that the patient has sustained and the real serious mischief that has in reality occurred, and which will eventually lead to the gravest consequences," Mr Erichsen goes on to say:—† "The shake or jar that is inflicted on the spine when a person jumping from a height of a few feet comes to the ground suddenly and heavily on his heels or in a sitting posture, has been well known to surgeons as a not uncommon cause of spinal weakness and debility. It is the same in railway accidents; the shock to which the patient is subjected being followed by a train of slowly-progressive symptoms indicative of concussion and subsequent irritation and inflammation of the cord and its membranes." This is a statement of the very deepest gravity, for it implies that injuries, in no wise directly affecting the spinal column, may yet give rise to inflammation of a most serious nature to the cord and its membranes. The implication is confirmed by the passage which immediately follows: ‡ "It is not only true that the spinal cord may be indirectly injured in this way, and that sudden shocks applied to the body are liable to be followed by the train of evil consequences that we are now discussing, but I may even go farther, and say that these symptoms of spinal concussion seldom occur when a serious injury has been inflicted on one of the limbs, unless the

* Op. cit., p. 155, Lecture VII, parts 1 and 2.
† Op. cit., p. 155.
‡ Op. cit., p. 156.

spine itself has at the same time been severely and directly struck. A person who by any of the accidents of civil life meets with an injury by which one of the limbs is fractured or is dislocated, necessarily sustains a very severe shock, but it is a very rare thing indeed to find that the spinal cord or the brain has been injuriously influenced by this shock that has been impressed on the body."

We all know that the symptoms of "shock" either pass away after such injuries, or that the "shock" may end fatally; but how comes it that after railway collisions, when fracture of limbs has occurred, these symptoms of "spinal concussion" are seldom seen, even though there may have been severe surgical collapse or "shock" as the result of the injury, and from which the patient has rallied in the usual way? Mr Erichsen tells us: *" It would appear as if the violence of the shock expended itself in the production of the fracture or dislocation, and that a jar of the more delicate nervous structures is thus avoided. I may give," he says, " a familiar illustration of this from an injury to a watch by falling on the ground. A watchmaker once told me that if the glass was broken, the works were rarely damaged; if the glass escapes unbroken, the jar of the fall will usually be found to have stopped the movement." The phenomenon of the watch is a matter of common observation, but we doubt the force of the analogy, unless, indeed, it can be shown that the watch has a nervous system or that is a sentient organism like ourselves. The statement is of itself sufficient, we think, to arouse scepticism as to the existence of such a condition as change in the spinal cord being the result of a blow altogether distant from it, or of a general shock upon the whole body; and the doubt is strengthened when we read that sleep has the same protective effect as fracture; for Mr Erichsen writes: "Those who are asleep at the time of the accident very commonly escape concussion of the

* Op. cit., p. 156.

nervous system. They may, of course, suffer from direct and possibly from fatal injury to the head or trunk; but the shock or jar, that peculiar vibratory thrill of the nervous system arising from the concussion of the accident is frequently not observed in them, whilst their more wakeful and less fortunate fellow-travellers may have suffered severely in this respect."* We ourselves know of no single case in which primary, and later secondary, changes have been certainly produced in the spinal cord, and in which there has not at the same time been some very clear evidence of injury on or close to the vertebral column; and we believe that a much simpler and more reasonable explanation may be offered of this immunity of the nervous system from any ill effects of concussion from general shock when a bone is broken than has here been advanced by Mr Erichsen.

We may fully agree with the author that it is impossible accurately to explain the change which has been produced in the spinal cord by concussion, just as it is difficult to explain how a heavy blow with the hammer deprives the magnet of its magnetic power. We must be content at present with the fact that there is a change, if indeed a fact it be. † "But," he goes on immediately to say, "whatever may be the nature of the primary change that is produced in the spinal cord by a concussion, the secondary effects are clearly of an inflammatory character, and are identical with those phenomena that have been described by Ollivier, Abercrombie and others, *as dependent on chronic meningitis of the cord and sub-acute myelitis.*"

"There is great variation in the period at which the more serious, persistent, and positive symptoms of spinal lesion begin to develop themselves. In some cases they do so immediately after the occurrence of the injury, in others not until several weeks, I might perhaps even say months had elapsed. But during the whole of this

* Op. cit., p. 120.   † Op. cit., p. 157. Italics our own.

interval, whether it be of short or of long duration, it will be observed that the sufferer's condition, mentally and bodily, has undergone a change." On this point the author " particularly insists." The man never completely gets over the effects of the accident. There may have been improvement, there has not been recovery. *"There is a continuous chain of broken or ill-health, between the time of the occurrence of the accident and the development of the more serious symptoms. It is this that enables the surgeon to connect the two in the relation of cause and effect. This is not peculiar to railway injuries, but it occurs in all cases of progressive paralysis after spinal concussion. . . . The friends remark, and the patient feels that 'he is not the man he was.' He has lost bodily energy, mental capacity, business aptitude. He looks ill and worn; often becomes irritable and easily fatigued. He still believes that he has sustained no serious or permanent hurt, tries to return to his business, finds that he cannot apply himself to it, takes rest, seeks change of air and scene, undergoes medical treatment of various kinds, but finds all of no avail. His symptoms become progressively more and more confirmed, and at last he resigns himself to the conviction that he has sustained a more serious bodily injury than he had at first believed, and one that has, in some way or other, broken down his nervous power, and has wrought the change of converting a man of mental energy and of active business habits into a valetudinarian, a hypochondriac, or a hysterical paralytic, utterly unable to attend to the ordinary duties of life." Truly this is a sad picture! And what tends to make the story even more unhappy is the fact, recorded just before as a remarkable phenomenon attendant upon this class of cases,† " that at the time of the occurrence of the injury the sufferer is usually quite unconscious that any serious accident has

* Op. cit., p. 158, *et seq.* † Op. cit., p. 157.

happened to him. He feels that he has been violently jolted and shaken, he is perhaps somewhat giddy and confused, but he finds no bones broken, merely some superficial bruises or cuts on the head or legs, perhaps even no evidence whatever of external injury. He congratulates himself upon his escape from the imminent peril to which he has been exposed. He becomes unusually calm and self-possessed; assists his less fortunate fellow-sufferers; occupies himself perhaps actively in this way for several hours, and then proceeds on his journey. When he reaches his home the effects of the injury that he has sustained begin to manifest themselves. A revulsion of feeling takes place. He bursts into tears, becomes unusually talkative, and is excited. He cannot sleep, or if he does, he wakes up suddenly with a vague sense of alarm. The next day he complains of feeling shaken or bruised all over, as if he had been beaten, or had violently strained himself by exertion of an unusual kind. This stiff and strained feeling chiefly affects the muscles of the neck and loins, sometimes extending to those of the shoulder and thighs. After a time, which varies much in different cases, from a day or two to a week or more, he finds that he is unfit for exertion, and unable to attend to business. He now lays up, and perhaps for the first time seeks surgical assistance." No evidence as yet, be it marked, of any lesion of or injury to the spinal cord, or the parts immediately about it; but, nevertheless, this is * "a general sketch of the early history of most of these cases of ' concussion of the spine ' from railway accident," which are of such frequent occurrence, and whose later symptoms, when the man has become a valetudinarian, a hypochondriac, or a hysterical paralytic, are due to chronic meningitis of the cord and subacute myelitis.

But let us learn, in the next place, what these symptoms are. They are given in full detail by Mr Erichsen, who

* Op. cit., p. 158.

analyses them and arranges them in the order in which they will present themselves, on making a surgical examination of a patient afflicted therewith. The main divisions must be here sufficient. The symptoms consist essentially in changes, derangements, or confusion of the *countenance*, the *memory*, the *thoughts*, the *business aptitude*, the *temper*, the *sleep;* in abnormal sensations in the *head;* in derangements of the *organs of special sense*, that of *vision* being more important and more common than those of *hearing, taste, and smell;* in derangements of the *sense of touch;* in the *speech;* in the *attitude of the patients*, which is stiff and unbending ; in hypersensitiveness of parts of the spine ; in the patient's gait ; in modifications of *motor power and sensation;* in *coldness and wasting of one or more of the limbs;* in the *state and nutrition of the muscles*, in *diminished electric irritability* of the muscles; in the weight of the body ; in the *genito-urinary organs*, in the *sexual desire and power*, and in the *pulse*.

* "*The order of the progressive development* of the various symptoms . . . is a matter of great interest in these cases. As a rule, each separate symptom comes on very gradually and insidiously. It usually extends over a lengthened period. In the early stages the chief complaint is a sensation of lassitude, weariness, and inability for mental and physical exertion. Then come the pains, tinglings, and numbness of the limbs ; next the fixed pain and rigidity of the spine ; then the mental confusion and signs of cerebral disturbance, and the affection of the organs of sense, the loss of motor power, and the peculiarity of gait." That the period of supervention of these symptoms should vary, that they should not all be always, or at the same time, present in every case, is only to be expected. The patient's early state may fluctuate. He may fondly hope to regain his health and strength, and more naturally so if the injury has been one of only "general

* Op. cit., p. 172, *et seq.*

nervous shock,"* or of "slight and indirect concussion of the cord," where "no immediate effects are produced, or if they are transitory," followed, commonly "after the first and immediate effects of the accident have passed off," by a "period of comparative ease, and of remission of the symptoms, but not of recovery." This fluctuating condition may go on for several weeks, possibly for two or three months, but "*there has never been an interval, however short, of complete restoration to health.*"† " So long as he is at rest he will feel tolerably well; but any attempt at ordinary exertion of body or mind brings back all the feelings and indications of nervous prostration and irritation so characteristic of these injuries; and to these will gradually be superadded those more serious symptoms . . . . . which evidently proceed from a chronic disease of the cord and its membranes. After a lapse of several months—from three to six—the patient will find that he is slowly but steadily becoming worse, and he then, perhaps for the first time, becomes aware of the serious and deep-seated injury that his nervous system has sustained."‡

"The chain of symptoms," continuous and unbroken, between the injury sustained and the illness subsequently developed, links the injury and the illness together in the relation of cause and effect. The railway accident has done its worst, the worst almost that it could do, for it has produced chronic meningitis of the cord, and subacute myelitis.

Now let us turn to the "pathology of concussion of the spine." We need hardly say it, for we all from our own experience know full well that the deadhouse is the only place where the pathological changes on which the symptoms of disease depend can be studied and learned; and that it is essentially from the teachings of the postmortem room that light has been cast in the present age of pathological activity and research on many of the dis-

* Op. cit., p. 173.   † Op. cit., p. 174.   ‡ Op. cit., p. 173.

eases about which little had been previously known. Without pathology, our knowledge of disease must be vague, and our treatment of it empirical and unsure; with it we pass from the realms of conjecture into those of comparative certainty and fact. If there be one thing which above all others has advanced our knowledge of disease, and made our treatment of it better than that of our forefathers, it has been the examinations of the dead, and of the morbid structures, which the post-mortem room has given to us, and which unquestionably make men like Wilks, Buzzard, Gowers, and Ross in our own country, and many others, far more trustworthy guides on diseases of the nervous system than Ollivier and Abercrombie, for whose teaching Mr Erichsen seems to have so strange a preference. "Nec silet mors" is the apt and solemn motto of the Pathological Society of London. Shall we not be astounded that "Mors silet" should face us in the very front of our inquiry into the pathology of concussion of the spine? Clinical facts have not been wanting, symptoms have been innumerable and grave, patients have been broken down, and men of mental energy have been changed into valetudinarians, hypochondriacs, and hysterical paralytics. What has the post-mortem room to tell us? Mors silet.

We pass over those graver lesions of which at the outset we have spoken in dealing with "concussion of the spinal cord," and of whose pathology already much is known, for it is only too often that death lays early claim to the victims of such injuries, and excludes thereby all necessity of medico-legal inquiry. Let us again follow Mr Erichsen. *"In those cases in which the shock to the system has been general and unconnected with any local and direct implication of the spinal column by external violence, and in which the symptoms, as just detailed, are less those of paralysis than of disordered nervous action, the pathologi-

* Op. cit., p. 175, *et seq.*

cal states on which these symptoms are dependent are of a more chronic and less directly obvious character." Then why, may we ask, should such cases be called "concussion of the spine," and why should the consideration of these lesser maladies be approached through an avenue of alarming detail as to the grave and fatal lesions which may immediately befall the spinal cord in the various accidents to which mankind is liable? Surely it would have been safer and wiser in writing of a class of injuries which hold so "important a position in medico-legal practice,"* to have separated and differentiated those symptoms which we must have recognised are much more cerebral or psychical, from those which can only find an explanation in some actual lesion of the spinal cord or of the nerves which are given off from it! All, however, are grouped together under the one common term "concussion of the spine," even though the patients may have only suffered from "vibratory jar," and there may have been "no evidence whatever of external injury."

"We should indeed be taking," the author goes on to say, "a very limited view of the pathology of concussion of the spine if we were to refer all the symptoms, primary and remote, to inflammatory conditions, either of the vertebral column, the sheaths of the spinal nerves, the meninges of the cord, or the substance of the medulla itself. Important and marked as may be the symptoms that are referable to such lesions as these, there are undoubtedly states, both local and constitutional, that are primarily dependent on molecular changes† in the cord itself, or on spinal

* Op. cit., p. 3.

† Nothing is of greater importance in any medico-legal inquiry than to have a definite meaning for the phrases which we may use. The word "changes" is so often used in speaking of these injuries to imply some condition underlying derangement of function, that it is essential to draw a very broad distinction indeed between "changes" which are obvious, visible, and gross, and those "changes" which are rather physiological and matter of pure conjecture. The act of

anæmia induced by the shock of the accident, acting either directly on the cord itself or indirectly, and at a later date, through the medium of the sympathetic, in consequence of which the blood distribution to the cord becomes disturbed and diminished." We may leave, at any rate for the present, the pathological condition spoken of as "spinal anæmia," for this is nothing more nor less than an assumption, or even, as Mr Erichsen says, *"a clinical expression possibly, more than a well-proved pathological fact."

More important, because certain and demonstrable if existent after death, are the lesions dependent upon inflammatory states. *" They doubtless consist mainly of chronic and subacute inflammation of the spinal membranes, and in chronic myelitis, with such changes in the structures of the cord as are the inevitable consequences of a long-continued chronic inflammatory condition developed by it." And yet "Mors silet." Let us hear the author's explanation thereof. †" It would at first sight appear a some-

writing these sentences is doubtless associated with some "change" in the cerebral cells and in the cells of the cord connected with the nerves of the arm. The "change" is not, however, a pathological state, and we do not know—there is, in fact, good ground for thinking otherwise—that the "changes" associated with purely functional derangement are of greater import than those which accompany such a movement as that to which reference has just been made. It is of course possible, though we know of no facts which prove, that functional derangement may, if long continued, become the means of perpetuating a vice, so to speak, in the presiding nerve-cells, just as bad habits may with great difficulty be broken or may even become master of the individual addicted to them. It may be remarked that were a portion even of the symptoms usually attributed to "spinal concussion" dependent on active organic disease, no sufferer could by any chance live through them, nor, indeed, would life be possible. The very magnitude and number of the symptoms are themselves powerful arguments—more powerful by the bedside than they may appear on paper—against the existence of undoubted *lesion* or pathological "change."

* Op. cit., p. 176.
† Op. cit., p. 176. Italics not in original.

what remarkable circumstance, that *notwithstanding the frequency of the occurrence of cases of concussion of the spine in railway and other accidents,* there should be so few instances on record of examinations of the cord after death in these cases. But this feeling of surprise will be lessened when we reflect on the general history of these cases. If in these, as in cases of direct injury of the spine with fracture or dislocation, the effects were immediate, severe, and often speedily fatal, surgical literature would abound with the details of the *post-mortem* appearances presented by them, as it does with those of the more direct injuries just alluded to. But, as in these cases of spinal concussion, the symptoms are remarkably slow in their development and chronic in their progress—as the patient will live for years in a semi-paralysed state, during which time the original cause of his sufferings has almost been forgotten—as he seldom becomes the inmate of a hospital—for the chronic and incurable nature of his ailments does not render him so much an object for such a charity as for some asylum or for private benevolence—and as the cause of his death does not become the subject of investigation before a coroner's court, there is little opportunity, reason, or excuse for a *post-mortem* investigation of that structure, which is probably the one that is least frequently examined in the deadhouse, viz. the spinal cord, as it is the one the correct pathological investigation of which is attended by more difficulties than that of any other organ in the body. Hence it is that, as in most other chronic nervous diseases that are only remotely fatal—as in cases of hysteria, neuralgia, and in nine tenths of those of epilepsy, we have no opportunity of determining in cases of concussion of the spine very remotely* fatal, what the

---

* "Remotely!" Although Mr Erichsen writes at p. 311, "And though, as Ollivier has observed, such a patient may live for fifteen or twenty years in a broken state of health, *the probability is that he will die in three or four.*" Italics our own.

anatomy of the parts concerned would reveal of the real cause of the obscure and intricate symptoms presented during life. So rare are *post-mortem* examinations of these cases that no instance has occurred to me in hospital or in private practice in which I could obtain one ; and, with one exception, I can find no record in the Transactions of Societies or in the periodical literature of the day of any such instance." We have, however, heard from Mr Erichsen himself wherein these pathological states consist, viz. in "chronic and subacute inflammation of the spinal membranes, and in chronic myelitis, with such changes in the structure of the cord as are the inevitable consequences of a long-continued chronic inflammatory condition developed by it "* and " Case 12 " is actually recorded as an example of this mode of death. Can any man believe that if these pathological states have been frequently induced by railway accidents, especially in those cases so very numerous, as we know they are, where there has been no blow on the spine itself and the injury has been only that of general shock or vibratory jar—can any man, we repeat, believe that amongst the many thousands who have up to this date been hurt in railway accidents, there should have been, *with one or two exceptions only*, no opportunity whatever of examining the spinal cord? For, be it observed, the cases of hysteria, neuralgia, and epilepsy, to which Mr Erichsen refers, bear no resemblance, have no analogy whatever, to these cases where he tells us there is chronic and subacute inflammation of the spinal membranes and chronic myelitis. How can Mr Erichsen reconcile this analogy with his own preceding statement that "the intraspinal inflammations," as he there very rightly says, " whether they affect the membranes of the cord, the cord itself, or both, are well recognised and easily determinable pathological states, the conditions connected with which are positive organic lesions that lie at the

* Op. cit., p. 176.

bottom of the functional disturbance?"* Analogous to them, and very strictly analogous too, are cases of tabes dorsalis, of lateral sclerosis, of insular sclerosis, of anterior poliomyelitis, and the whole host of diseases of the spinal cord—" chronic and incurable" and "remotely fatal"— which the great pathologists of the day have by examinations of the spinal cord for years been making known to us. How comes it that while these and kindred diseases are eagerly watched and investigated by our pathologists, the chronic and incurable diseases produced by railway accidents should remain altogether unobserved, and that no cases have been placed on record to tell what is the pathological basis of the symptoms? How is it that these diseases of the spinal cord are, if we mistake not, rarely or never seen in the special hospitals devoted to them, and that at such a hospital as the National Hospital for Epilepsy and Paralysis, in Queen Square, with its justly distinguished staff, they should be well-nigh unknown? To our mind it is absolutely inconceivable that had an organ so important in the animal economy as the spinal cord—an organ, moreover, singularly liable not to recover from injuries inflicted upon it, or from diseases attacking it†

* Op. cit., p. 176.

† As has been well pointed out in an important paper by Dr Long Fox ('Lancet,' vol. i, 1882, p. 6), "Note on the Curability of Tabes Dorsalis," where he records a case in which there was a complete remission of some of the gravest phenomena of the disease after they had existed for four years, and yet on death from accidental causes the usual pathological changes were found in the spinal cord. Dr Fox writes, " Have we as yet any proof that sclerosis can be recovered from? The remission of symptoms, even for several years, leads to the belief that under treatment directed to the tone of the vessels, or the nutrition of the cord, other tracts of this organ take on the duties that we consider the attributes of the posterior root zones. But during this very remission, if we may judge by pathological anatomy, the lesion is slowly, though surely, following a progressive course; and, although there is nothing impossible in the more hopeful view of cure, I would still repeat that so far published facts are wanting to prove

—been frequently damaged or affected in the way Mr Erichsen surmises, and with thousands of sufferers in railway collisions to afford us examples during many years, there should have been, with two or three solitary exceptions, no *post-mortem* investigations to guide us to a real and unquestionable knowledge of the pathology of the symptoms to which the injuries have given rise.

But, on the other hand, if there be this great difficulty, nay, impossibility, of making *post-mortem* examinations in these cases, where, let us ask, are those who are living in semi-paralysed states, *not* forgetful, depend upon it, of the cause of their sufferings; and if from the nature of their ailments not in our hospitals, objects yet for the care of an asylum or for private benevolence? Where are they, for they must be numbered by scores upon scores? And this other question we must also ask before we can accept Mr Erichsen's teaching upon the subject, how many injured persons have there not been who have had all or most of the symptoms which he has detailed and described, and whose spinal cords and spinal membranes have been undergoing these supposed inflammatory changes, who have recovered, and that, too, with unexpected speed, when the termination of any medico-legal inquiry or litigation has removed—as we shall subsequently have to point out—an intolerable burthen from their minds?

After dwelling briefly upon the difficulty there may be in determining during life what parts of the spinal cord are affected, and upon the important fact that spinal meningitis and myelitis most frequently coexist, the symptoms of meningitis predominating in one case, those of myelitis in another, the " characteristic appearances after death presenting a predominance corresponding to that it, whilst, on the other hand, many cases of temporary improvement of symptoms have shown eventually post mortem a steady progress of the lesion."

assumed by their effects during life;"* Mr Erichsen goes on to say: "I have given but a very brief sketch of the pathological appearances that are usually met with in spinal meningitis and myelitis, as it is not my intention in these lectures to occupy your attention with an elaborate inquiry into the pathology of these affections, but rather to consider them in their clinical relations. I wish now to direct your attention to the symptoms that are admitted by all writers on diseases of the nervous system to be connected with and dependent upon the pathological conditions that I have just detailed to you, and to direct your attention to a comparison between these symptoms and those that are described in the various cases that I have detailed to you as characteristic of 'concussion of the spine' from slight injuries and general shocks of the body. The symptoms that I have detailed at pp. 156 to 175 arrange themselves in three groups:—1st. The cerebral symptoms ; 2nd, the spinal symptoms ; 3rd, those referable to the limbs." The symptoms under these respective headings are then compared with the symptoms described by Abercrombie and Ollivier, in the fatal cases which they record of cerebro-spinal meningitis and myelitis, and Mr Erichsen concludes: †"If we take any one symptom that enters into the composition of these various groups, we shall find that it is more or less common to various forms of disease of the nervous system. But if we compare the groups of symptoms that have just been detailed, their progressive development and indefinite continuance, with those which are described by Ollivier and other writers of acknowledged authority on diseases of the nervous system, as characteristic of spinal meningitis and myelitis, we shall find that they mostly correspond with one another in every particular—so closely, indeed, *as to leave no doubt that the whole train of nervous phenomena arising from shakes and jars of*

* Op. cit., p. 183.
† Op. cit., p. 187. Italics our own.

or blows on the body, and described at pp. 156 *to* 175 *as characteristic of so-called* 'concussion of the spine,' *are in reality due to chronic inflammation of the spinal membranes and cord.* The variation in different cases being referable partly to whether meningitis or myelitis predominates, and in a great measure to the exact situation and extent of the intra-spinal inflammation, and to the degree to which its resulting structural changes may have developed themselves in the membranes or cord."

To the train of nervous phenomena described at pp. 156 to 175 we have already referred, and they may be found in those parts which we have quoted from the author, commencing with the seductive analogy of the watch, and ending with the deep-seated injury that the nervous system has sustained. It is thus consecutively laid down for our acceptance, as Mr Erichsen's teaching on this most important subject, that the symptoms occurring too frequently "in groups, sometimes of scores at a time,"* are due to chronic inflammation of the spinal membranes and the cord. It will be an indirect object of this work to show that, *with very rarest exception,* the spinal cord is absolutely uninjured in these cases of railway collision, shock, or jar, and that now, not less than of old, the spinal cord maintains its supremacy as the most securely protected of all the organs of the body.

It is necessary that we should now return to the only case on record with which Mr Erichsen is acquainted, and the post-mortem account of which we shall take direct from the 'Transactions of the Pathological Society.' The clinical history is supplied by the author as having been obtained from Mr Gore, of Bath.† "The patient was a middle-aged man, 52 at the time of death, of active business habits. He had been in a railway collision, and, without any sign of external injury, fracture, dislocation, wound, or bruise, began to manifest the usual nervous symptoms.

* Op. cit., p. 7.   † Op. cit., p. 178.

He very gradually became partially paralysed in the lower extremities, and died three years and a half after the accident. Immediately after the collision the patient walked from the train to the station close at hand. He had received no external sign of injury, no contusions or wounds, but he complained of a pain in his back. Being most unwilling to give in, he made every effort to get about in his business, and did so for a short time after the accident, though with much distress. Numbness, and a want of power in the muscles of the lower limbs gradually but steadily increasing, he soon became disabled. His gait became unsteady, like that of a half-intoxicated person. There was great sensitiveness to external impressions, so that a shock against a table or chair caused great distress. As the patient was not under Mr Gore's care from the first, and as he only saw the case for the first time about a year after the accident, and then at intervals up to the time of death, he has not been able to inform me of the precise time when the paralytic symptoms appeared; but he says that this was certainly within less than a year of the time of the occurrence of the accident. In the latter part of his illness some weakness of the upper extremities became apparent, so that if the patient was off his guard a cup or a glass would slip from his fingers. He could barely walk with the aid of two sticks, and at last was confined to his bed. His voice became thick, and his articulation imperfect. There was no paralysis of the sphincter of the bladder until about eighteen months before his death, when the urine became pale and alkaline with muco-purulent deposits. In this case the symptoms were in some respects not so severe as usual, *there was no very marked tenderness or rigidity of the spine,** nor were there any convulsive movements." There was no apparent difficulty in making a post-mortem examination of this case, and the

---

* Italics our own.

cord was examined by Dr Lockhart Clarke, whose report was as follows :—

* "*Diseases of the brain and spinal cord consequent on a railway collision.*—This, I believe, is the first of these curious cases in which the condition of the brain and spinal cord has been ascertained, or at least recorded. The patient was under the care of Mr Gore, of Bath, who sent me the spinal cord, together with the following particulars :—' A gentleman, aged 52 at the time of death, had been the subject of many distressing symptoms, all arising out of the shock he received in a railway collision three and a half years before his death. He had no wounds, nor fractures, nor material contusions ; but having been previously an active, intelligent man, conducting with success a large business, he began at once to suffer vaguely with pains down the back and in the head, though not of a very acute kind. He gradually, though very slowly, failed in every respect as to mind and body—not, however, losing his intellect. He became unable to walk with steadiness or firmness, and before his death, which was clearly hastened—say three months—by an accidental fall, he could barely walk with the aid of two sticks, and was for the last month confined to his bed. One other symptom of disorder, clearly connected with all the others, was inability of the bladder, with eventually want of control over it. The urine also was pale, turbid, and alkaline, with muco-purulent deposits. Nothing of the kind had existed before.

" At the *post-mortem* examination, in addition to a generally shrunken and wasted condition of the spinal cord, there was in the brain, general, though slight, opacity of the arachnoid, with sub-arachnoid effusion. The cortical substance of the brain generally was pallid and soft ; this was particularly the case on the under surface of the

* 'Transactions of the Pathological Society of London,' 1866, vol. xvii, p. 21, *et seq.*

anterior lobes on both sides. (His speech had been for a month or six weeks thick and hesitating.) The kidneys were much disordered, hard, dense, and with many isolated purulent deposits. The bladder was contracted, and its mucous membrane pulpy and vascular.

"On examining the spinal cord, as it was sent to me by Mr Gore, I found that the membranes at some parts were thickened, and adherent at others to the surface of the white columns. In the cord itself, one of the most striking changes consisted in a diminution of the antero-posterior diameter, which, in many places, was not more than equal to half the transverse. This was particularly the case in the upper portion of the cervical enlargement, where the cord was consequently much flattened from behind forward. On making sections, I was surprised to find that of all the *white* columns, the *posterior* were exclusively the seat of disease. These columns were darker, browner, denser, and more opaque than the antero-lateral; and when they were examined, both transversely and longitudinally, in their preparations under the microscope, this appearance was found to be due to a multitude of compound granular corpuscles, and isolated granules, and to an exuberance of wavy fibrous-tissue disposed in a longitudinal direction. It was very evident that many of the nerve-fibres had been replaced by this tissue, and that at certain spots or tracts, which were more transparent than others, especially along the sides of the posterior median fissures, they had wholly disappeared. Corpora amylacea also were thickly interspersed through the same columns, particularly near the central line. The extremities of the posterior horns contained an abundance of isolated granules like those in the columns; and in some sections the transverse commissure was somewhat damaged by disintegration. The anterior cornua were decidedly smaller than natural, and altered in shape, but no change in structure was observed.

"The striking resemblance between this case and cases

of locomotor ataxy, as regards the limitation of the lesion of the white substance of the cord to the posterior columns, although the *nature* of the lesion is somewhat different, excited my curiosity whether the paralysis and difficulty of locomotion partook of the nature of ataxy. In a letter to Mr Gore I inquired whether the patient's gait was remarkable for its unsteadiness, like that of a man somewhat intoxicated; or whether the movements were jerking or spasmodic. In reply I received the following information :—' The semi-paralytic state which I attempted to describe was precisely that of unsteadiness, somewhat like that of partial intoxication ; but, on the other hand, there was very little, if any, jerking or twitching.' "

It will hardly surprise us to learn that objections have been raised against receiving this case as an unequivocal instance of injury to the spinal cord by general concussion, shock or jar. The case is given in detail by Mr Shaw in his article on " Injuries of the Back " in ' Holmes's System* of Surgery,' and he there writes : " The first remark which the reading of the above case suggests, is concerning the disproportion that appears between the slight injury sustained by the patient and the magnitude of the results, occupying three and a half years in coming to an end. From progressive paraplegia, and a morbid condition of the spinal cord, identical with what has been described in the case, being known to originate independently of injury of any kind, the question forces itself upon us, was the shock in the railway accident a cause, or a coincidence ? This doubt would not have been expressed if Mr Gore had been in attendance on the gentleman at first; but, according to the narrative, he did not see him till a year after the accident.

" But granting, for argument's sake, that a shock, analogous to concussion of the brain, had really been received, the question may be asked: how would that assist in

* Second edition, vol. ii, p. 377, *et seq.*

accounting for the peculiar morbid change found in the cord on dissection? The concussion would be followed by inflammation. But that inflammation would be general: it would extend over the whole surface, and enter into the deep structures of the cord, including every column equally. The organic change, however, is not general but partial; it is confined to the posterior columns. How, then, is that selection or limitation to be explained?

" Another objection presents itself. According to modern views of pathology, the morbid action concerned in producing granular degeneration of the tissues is distinct from inflammation. The process is seen in operation in fatty degeneration of the muscular substance of the heart, in the formation of the arcus senilis, in the production of atheroma in the arteries, &c., and in none of these instances is the degeneration preceded by inflammation. It may be argued, therefore, that the organic changes in the columns of the spinal cord, consisting of degeneration of the nerve-fibres, depend on some cause not hitherto ascertained, different from inflammation; and that accordingly their connection with concussion of the cord is merely hypothetical." To these objections Mr Erichsen makes no reference. That they are weighty and well-nigh overwhelming we must allow.

In seeking for the *cause* of such morbid changes as were present in this case, we must not be unmindful of the readiness with which anything that lowers the general tone and strength may bring to light the existence of pathological states or diseases, by the development of symptoms which in the previous state of health had been in abeyance. Examples are seen every day in cases where lesion of the nervous centres has no part. A sprain or severe blow may be followed by an acute attack of gouty inflammation; the loss of strength which now and then accompanies parturition or prolonged suckling may reveal an hypermetropia which vigorous and lusty power of accommodation had been able to hide. And that the earliest symptoms of a progressive degeneration of the spinal cord should have been unobserved before, and should have been first made manifest after, a serious shock to the system, is not a circumstance to excite our great surprise. It is, moreover, well recognised that the symptoms of tabes may undergo a singular remission or abeyance, and give rise to the impression that the disease has been cured.*

On the other hand also, although there was in this case no " sign of external injury, fracture, dislocation, wound, or bruise," it is a point worthy of mention that the patient " complained of a pain in his back," as far as can be gathered from the imperfect early history, almost immediately after the collision. That this pain in the back may have indicated a much more serious injury to the vertebral column than physical signs led those who saw him to suspect, is another objection, it seems to us, to receiving this case as an undoubted instance of disease of the spinal cord due to a general shake or jar, apart from any direct injury to the spinal column or its contents. On no one of these objections, however, would we seek to lay undue stress, nor shall we deny that the case may have been precisely as it has been recorded by Mr Erichsen. We must, however, once more point out how extraordinary it is that this still remains the only case of the kind on record, and that against it important objections, made in no spirit of cavil or bias, have been raised.

Granting, however, to the full the difficulty there may be in obtaining post-mortem examinations in private houses, it is yet something more than remarkable that no further evidence exists of the pathological lesions in these cases of concussion of the spine, on the frequency of whose occurrence Mr Erichsen lays so much stress; and especially when we consider that the semi-paralysed sufferers must have been under medical care up to their last moments,

* See previous note (p. 96) and also note on p. 153.

and that many years have now elapsed since the attention of the whole profession was drawn to the subject by the publication of this case. Is it too much to say that had chronic meningitis and myelitis been present in but a fraction of the numbers which Mr Erichsen's writings would lead us to suppose, *" experiments made by disease " would have shed floods of light on the function of the spinal cord? But, we ask, what have railway collisions done to advance our knowledge of the physiology and pathology of the spinal cord, and of the nerves which are offshoots from it? The answer is, Nothing. *Ex nihilo nihil fit;* nor are we justified in yet believing, from the evidence which has been adduced, that the nervous phenomena arising from shakes and jars of, or blows upon the body, and described as characteristic of " concussion of the spine " are, in reality, due to chronic inflammation of the spinal membranes or cord, or that they are even due to any pathological lesion of the spinal cord at all.

Having thus far devoted our attention to the views of Mr Erichsen, let us glance, in conclusion, at the opinions of other surgical authorities on " concussion of the spine." Mr Shaw† writes as follows:—" The term concussion as applied to the spinal cord has obviously been derived from a supposed analogy between the injuries occurring to it and to its kindred organ the brain . . . .
. . . . Certain cases of injury of the back are met with in which paraplegia has directly or shortly afterwards occurred, and in which, upon examination of the spine after death, no fracture, displacement, extravasated blood, or anything capable of compressing the cord, can be discovered. The explanation therefore given is that the spinal cord has been damaged by concussion."

* A phrase first used, if we mistake not, by Dr Hughlings Jackson, who, by a careful study of the symptoms or physiological disturbances of its localised diseases, has done so much to elucidate the functions of the brain.

† 'Holmes' System of Surgery,' 2nd ed., vol. ii, p. 371, *et seq.*

He then quotes a case from the 'British Medical Journal' of April 24th, 1869, where a man had fallen from a van, and lighted on his head and shoulders. There was instant paralysis of both upper and lower extremities, and he died in sixty hours. "On examination, neither fracture nor dislocation was found in any part of the spine, nor was blood effused in the canal. But, on making a section of the cord opposite the third and fourth cervical vertebræ, a clot of blood was found lying in its centre." "From the remarks appended to the case by the narrator," he goes on to say, "it is obvious that he had no doubt of its having been a case of concussion, and it would seem that the same opinion would be held generally. But it is evident that the lesion in the cord admitted of being satisfactorily accounted for otherwise:" by "sudden violent and extreme bend or twist of the flexible cervical region of the spine," or by sharp and acute flexion of the cord itself, whereby a blood-vessel was ruptured in its interior.* After some remarks upon the greatly increased interest there now is in spinal injuries in consequence of the "numerous formidable evils that result from accidents in railway collisions," the writer proceeds :—"Why the term 'concussion'

* More remarkable even than this case in its strange misconception, as it seems to us, of the cause, is one published in the 'British Medical Journal' of November 6th, 1880, by Dr W. J. H. Lush, and entitled, "A Case of Infantile Paralysis." A child of three, in perfect health, was jolted in its perambulator, "*the concussion being so great as to throw the occupant of the carriage violently forward and immediately to jerk the body back again.*" The child began to cry directly, saying, "Oh, my back," &c. It passed a restless night, and the next morning was found to be "completely paraplegic, though there was little, if any, loss of sensation." After some valuable remarks on the importance of observing the earliest symptoms of illness in a child, and of not assuming that it is only cross, the author adds, "In this particular case, in the absence of any other cause, I think it may with justice be said that the paralysis was due to concussion of the spinal cord, the result of the sudden shock." (Italics our own in this quotation.)

should have been applied to an injury of the brain, is not difficult to comprehend; but the same cannot be said in reference to the spinal cord," because the physical surroundings of the two organs are totally different. Between the cranium and the brain no "dampers" are anywhere interposed, and the shake or concussion must be general; while the spinal cord is surrounded by dampers, and occupies the centre of a "roomy chamber" at a considerable distance from its osseous walls. "Even if the spine did vibrate, there would be no connecting medium capable of conducting the vibrations to the cord." And then, speaking of the effects "produced on the nervous system by the injuries chiefly characteristic of serious railway accidents, it does not appear," he says, "that the term 'concussion of the spine,' so often used in connection with them, has been well chosen. It ought to have been of more comprehensive signification," for in a violent collision "the whole body is in a concussion." . . . "When the body is jolted, jarred, and shaken in the violent manner described, . . . it is self-evident that the viscera will be stirred and jolted likewise; that their internal structure will be subject to be bruised or lacerated, giving rise to ecchymosis; and that the nerves on their passage may be stretched, torn, or otherwise injured. Thus, all the great constituent parts of the nervous system, brain, spinal cord, and sympathetic system, are included in the common risk of the catastrophe. And the account would be incomplete, if the influences of mental shock—that of fright, and of witnessing appalling spectacles—were neglected."*
He then refers to the progressive paraplegia which has been "represented as a sequel of what is called 'concussion of the spine,'" recording, as an example, the case already given where the spinal cord was examined by Dr Lockhart Clarke, and he thus concludes: "On the whole, it may be affirmed, that what is most wanted for the better

* Op. cit., p. 374.

understanding of those cases commonly known under the title of 'concussion of the spine,' is a greatly enlarged number of post-mortem examinations. Hitherto our experience has been derived almost wholly from litigated cases, deformed by contradictory statements and opinions; and the verdicts of juries have stood in the place of post-mortem reports."*

In his 'Lectures on the Principles of Surgical Diagnosis,'† Le Gros Clark deals at considerable length, and obviously after large experience, with the whole subject of "concussion of the spine," whether met with after railway collisions or ordinary accidents. Speaking more especially of that ‡ "class of cases the most numerous where there is no evidence of physical lesion, and in which the symptoms supervene at an interval, longer or shorter, after the occurrence of the shock," he says, "in some which are characterised by loss of muscular strength, shrunk limbs, defective power in co-ordination, feeble exercise of motive volition, and even paralysis and imbecility, there must be organic change which is often progressive. . . . . But in others the morbid condition is evinced by deteriorated health and defective nerve energy; and in these probably the impaired health is due to the indirect influence exercised on the organs of assimilation, and a reactionary impression exerted on the nerve-centres themselves. . . . . The peculiarity of these accidents appears to be that the spine is roughly jarred, sometimes subjected to a succession of minor concussions or violent oscillations, and that the entire nervous system is simultaneously agitated. There is, consequently, no immediate persistent incapacity, as in ordinary concussion from a blow on the head or spine; but a prolonged series of symptoms ensue, directly traceable to the shock, and often assuming a more aggravated character as time elapses."

* Op. cit., p. 378.   † Churchill, 1870.   ‡ P. 70.

"Simple* *concussion* of the spine occurs as the consequence, usually, of a fall on the nates or back. I say concussion, because the analogy in the symptoms which characterises this condition and so-called concussion of the brain, justifies the use of the expression." He then records two cases where there was complete paraplegia from falls of this kind, passing away in the course of a few months, and he says, "I cannot dismiss from my mind the impression, that in these and similar protracted cases, there is something more than simple concussion needed to account for the duration of the symptoms; probably extravasation of blood in the theca or canal, which is slowly absorbed." And he regards† "hæmorrhage as the probable cause of the protracted symptoms" in a case of paraplegia coming on a few hours after a man had been injured in a railway collision. The patient recovered slowly but perfectly. "One spot on the back was always tender, and continued so still at times."

But of the "special form" of railway concussion he says :—"I speak of it as special because there certainly are distinctive characteristics attending railway concussion of the spine, which are exceptional, to say the least, in similar injuries otherwise produced. And this exceptional character consists in the curiously diversified results which are met with; sequences which seem to be more allied with general nervous shock, and consequent deteriorated innervation, than upon special shock or concussion of the spinal cord. Indeed, it is difficult to explain many of the sequelæ of these injuries without supposing that the organic nerves, and therefore functions, are seriously implicated; that the excretory functions are imperfectly performed; that the organic chemistry is deranged, and that the source of life, the blood, is poisoned. . . . . The sequelæ are more varied and protracted than in ordinary concussion; a circumstance

* Op. cit., p. 145. † Op. cit., p. 148.

which is probably explained, in a measure, by the influence of emotion." . . . . "I* think it not inconsistent with acknowledged facts, to affirm that protracted functional disturbance, or even fatal disease, may be the consequence of a rude shock, simultaneously, to the nerve-centres of the emotions, of organic and of animal life. I am therefore disposed to regard these cases of so-called railway spinal concussion, as, *generally, instances of universal nervous shock, rather than of special injury to the spinal cord.*† At the same time I admit that in this class of cases we meet with instances of simple concussion, but I see no reason for taking *them* out of the category of concussion of the spine from other causes."

If we turn to 'Ziemssen's Cyclopædia,' where Prof. Erb has clearly brought before his readers all the information which the literature of the subject has been able to afford, we find the following judicial estimate of the nature of "concussion of the spinal cord :"‡—"We include those cases in which energetic traumatic influences (falls, blows, collision, &c.) have given rise to *severe disturbances of the function of the cord, without any considerable visible anatomical changes in the latter.* Slight changes, small capillary extravasations, &c., probably exist in such cases, but they do not seem to constitute the proper essence of the disease; for the most part the anatomical change is quite negative, and we do not yet know what changes, if any, constitute the basis of the concussion proper. . . . . The diagnosis is in many cases so uncertain, and the want of satisfactory evidence from autopsies so great, that the history of the disease is still surrounded by darkness. .
. . . It is therefore, rather rash to entertain a decided opinion regarding the *proper nature of concussion of the cord.* It seems to be certain that the anatomical report is a negative one. The most common view, therefore, is that

* Op. cit., p. 151. † Italics our own.
‡ Vol. xiii, English translation, p. 344, *et seq.*

which supposes only molecular changes in the finer nerve-elements to have occurred, giving rise either to an immediate and complete functional paralysis of the latter, or forming the commencement of further disturbances of nutrition, which at a later time may result in degenerative inflammation. . . . . Much remains to be done in this respect" (*i. e.* in the more correct estimate and clearer definition of cases); "the first thing consists in collecting accurate reports of cases, avoiding, more carefully than has hitherto been done, the intermixture of other sorts of lesions."

To these remarks we need add but little. A study both of concussion of the spinal cord and of "concussion of the spine," be they the same thing or be they something different, makes it very clear that lesion of the spinal cord from simple concussion blow is very rare indeed, and that the existence of meningo-myelitis—an easily-recognised pathological condition—as a remote or early consequence of some vibratory effect upon the cord, still lacks the solid basis of established observation.

CHAPTER III

THE COMMON SPINAL INJURIES OF RAILWAY COLLISIONS

It will at once appear obvious how necessary it has been to deal thus fully in the two preceding chapters with the clinical and pathological aspects of concussion of the spinal cord and concussion of the spine, when we point out that injuries of the back more often become the subject of medico-legal inquiry than other kinds of injury to which man is liable. So frequently is this the case that of 250* consecutive cases of all kinds where there was medico-legal investigation, we find no fewer than 145† complained of their backs or of their "spines."

We will speak in the first place of the more common injuries of the back received in railway collisions, and then afterwards consider the individual cases, for they are very few, where there has been some undoubted injury to the nerve elements of the spinal column.

We premise that we are dealing with perfectly genuine cases, where none of the circumstances, as far as we were

* A number not chosen for statistical purposes, but simply as a matter of convenience, 250 cases filling a note-book.

† The same observation has been made by J. Rigler, 'Ueber die Folgen der Verletzungen auf Eisenbahnen, insbesondere der Verletzungen des Rückenmarks,' Berlin, 1879. Rigler gives statistics which show that since the passing of a law in Germany for the compensation of persons injured on railways the number of injuries or complaints of injuries had enormously increased, and that, moreover, of thirty-six complaints after injury no fewer than twenty-eight were of the back.

able to learn, threw doubt on the *bona fides* of the patients.

Let us look at a very common occurrence. The patient has been in a collision; he was perfectly conscious at the time that he met with no blow—knows, in fact, exactly what occurred to him when the accident happened, and yet he finds that within a few hours, occasionally much sooner, he is seized with pain in his back. What has happened to him? We will examine a case where the symptoms are simple and unobscured by any other injury.

M. A—, a strong and active man, was riding in a first-class carriage, when a slight collision took place. He was, at the moment, leaning forwards reading, and was not even moved from his seat. He felt a little upset and shaken, and had some brandy in consequence, but he was able in a few minutes to set off and walk to his business. The next day he felt some pain in the lumbo-sacral region, which on the following day became acute, especially on movement, and on the third and fourth days after confined him to the house. He was ordered a belladonna plaster, and in a week he began to improve, though having occasionally sharp pain. There was no local tenderness.*

It is clear from this history that the injury was a simple sprain of the muscles and ligaments about the lumbo-sacral region. It was, in fact, a "traumatic lumbago." A slight collision, such as this man was in, has the tendency to throw the body suddenly backwards or forwards, and, although in this case the patient had not been moved, the momentum† is usually sufficient to throw or jerk the traveller from his seat. Unconscious effort is probably made at the instant to hold the back rigid, and we find as

---

* See also Cases 132, 170, 234, Appendix, and many others.

† In a series of papers "On Cases of Injury from Railway Accidents," 'Lancet,' vol. i, 1867, p. 389, *et seq.*, Dr Buzzard touched on the *dynamics* of railway collisions. The subject is an important one.

a result of the violence and of the sudden resistance induced by "setting" of the muscles and ligaments, that the ligaments are stretched, and the muscular attachments are likewise strained in the dorso-lumbar or lumbo-sacral regions of the column. The injury is precisely the same as that which we meet with in everyday practice, where a man complains to you that while lifting a heavy weight (see Case 234, Appendix) he suddenly felt a severe and acute pain, which almost prevented him from moving, in the lower part of his back. "This sudden pain is probably caused by cramp or the rupture of some fibres of a muscle during the act of contraction."* You examine him and can find no external sign of injury to the back, but he hesitates to stoop when you ask him, he holds his back unnaturally stiff, he finds it difficult or impossible to rise from his seat, and very likely there is some local tenderness in the muscular mass on either side of the lumbar vertebræ. The case recited is one of the most common and most simple kinds, but we may meet with the same injury in very different degrees of severity. Let us take another example:—

A rather smart collision caught a man, æt. 58, sitting upright in the carriage with his head slightly turned to one side. He was thrown back, and his head was knocked against the carriage, the brim of his hat fortunately saving him from a severer blow. He felt shaken and sick, but did not vomit. Within a couple of hours of the accident he was seized with pain, and *tenderness* was felt in the lower part of the back, especially over the two lower dorsal and two upper lumbar vertebræ. He was taken home and put to bed, where he lay for a month suffering at first from such severe pain *throughout the whole spine*—cervical, dorsal, lumbar, and sacral regions— that he was barely able to move. There was never any

* See lecture on "Backache," by Dr Geo. Johnson, F.R.S., 'British Medical Journal,' vol. i, 1881, p. 222.

acceleration of pulse, elevation of temperature, or peripheral pain. At the end of the month he began to improve, and was able to move his arms and his head without pain, and occasionally to sit up in bed. In a couple of months he was able to get up, and in three months to move about so well as to do a little business. He gradually recovered, and is now, five years after the accident, in good health, though when travelling he still feels an "uncertainty," and unless he carefully supports himself is liable to have a return of pain in the lower part of the back.

These cases afford good examples of the same kind of injury, though at the two ends of the scale. It must not, however, be thought that it is usual to meet with cases such as these where there is no other complication. "Nervous shock" in its varied manifestations is so common after railway collisions, and the symptoms thereof play so prominent a part in all cases which become the subject of medico-legal inquiry, whether they be real or feigned,* that we are almost sure to meet with the symptoms of it associated with pains and points of tenderness along the vertebral spinous processes. From what we have ourselves seen, and from the arguments which we have often heard used about individual cases, we cannot help thinking that it is this combination of the symptoms of general nervous prostration or shock and pains in the back, such as these two cases presented, which has laid the foundation of the views—erroneous views as we hold

---

* Recently (see 'Times,' Saturday, February 19th, 1881) a case was recorded of an action against a railway company for injuries received by a man in accidentally putting his leg through a hole whereby he was thrown down and hurt his leg. The Judge commented strongly on the fact that the symptoms complained of bore a strange resemblance to those heard of so frequently after railway *collisions*. The symptoms, we think, in such cases ought properly to be termed "litigation symptoms," as we shall subsequently have to show.

them to be—so largely entertained of the nature of these common injuries of the back received in railway collisions.

A man shows signs of having received a nervous shock, and of these we shall speak at a later stage, and at the same time he is alarmed himself, and gives alarm to others by the pain and tenderness, often very severe, which he suffers at some part of the spinal column, or of the structures lying contiguous to it. In the first case recorded there can be no doubt that a simple muscular or ligamentous strain was the sole cause of the pain; nor can there be much doubt, if indeed there can be any, from the absence of every indication of other injury, that there was a like cause of the pain, extending throughout the whole spinal column, and making every movement of trunk and limb almost intolerable, which was so distressing in the second case. We do not meet with such cases in ordinary practice, but we see no reason why the whole trunk may not be so suddenly rocked and swayed from side to side and backwards and forwards that, with the free movement allowed to it in every direction by its many joints, the innumerable muscular and ligamentous attachments of the spinal column may not be strained and stretched; thus causing pain, severe in character, throughout almost the whole length of the back.* Railway collisions, however, provide the conditions which determine the possibility of such extensive strain of the vertebral column. Now one part, now another, is sprained in the jerks and jolts which accompany most collision accidents, and the pain, more commonly situated in the lumbar region alone,

* It is surgically interesting to note how rare it is to see any disease or inflammation of the small joints which are to be found between the vertebræ themselves or between the vertebræ and the ribs. "It may be deemed singular," Mr Shaw writes in his essay on "Disease of the Spine" ('Holmes's System,' vol. iv, p. 103), "that, numerous as are the small joints formed by the opposing surfaces of the oblique processes, in the posterior segment, disease is scarcely ever witnessed in them."

may thereupon affect other parts of the column. And very variable may be this pain both in range of distribution and in character.

In a case recently under our care a young man who was in a bad collision began three days afterwards to feel pain and stiffness across the loins, so that he moved with difficulty and felt easiest when flat upon his back. The pain in his back gradually increased, and, to use his own words, he thought all was over with him, and that he was going to be paralysed, for when he tried to stand up a sharp pain seized him in the back like a knife cutting into him, and shot downwards and upwards from the loins like an electric shock so that he dropped upon the floor. This state of things lasted for about three weeks, during which he had also the greatest difficulty in defecation and micturition. There was at the same time considerable local tenderness, but never any outward signs of injury. This description is almost characteristic of an attack of acute lumbago.

Hitherto we have spoken more especially of pain—pain, that is, upon movement either of the body or of the limbs, and have laid no stress upon the occurrence of tenderness at the same time. Does this tenderness, felt on pressure over one of the spinous processes, or even over the transverse processes of the spinal column, indicate graver injury than is signified by the pain which is more common and more extended in its area? We think not. Precisely the same tenderness may be met with, as we ourselves have seen, in cases of simple "sprain of the back," from lifting heavy weights; and it would be strange if, after the severe wrenching and straining which are likely to be sustained in a collision, we did not frequently meet with actual local tenderness as well as pain. In the severer cases the same condition is in all probability produced as we can obviously see in and around an ankle or other joint which has been sprained. There are pain and tenderness, and swelling with discoloration, and we only do not see the swelling about the small joints of the vertebral column and about the muscular and ligamentous attachments, because the structures sprained are more deeply situated, and are much smaller in size. A deep-seated discoloration, however, may be sometimes seen, even where there has been no blow. Is this tenderness, then, *by itself* a more dangerous symptom than the pain on movement? Let us ask if tenderness of the back is a prominent symptom in cases of spinal cord disease which we see in our hospital wards? Is it not rather conspicuously absent, and is not a diagnosis made by a totally different train of symptoms? No, tenderness by itself, not less than pain by itself, is not an indication of grave injury to the contents of the spinal column; it is a symptom which, if of any value at all, and we shall see as we proceed of how little value as a symptom tenderness of the back may be, is one rather to reassure, as pointing to the kind of injury with which we are here attempting to deal. "It is worth inquiring," writes Laycock,* " *what* we press, when we exert pressure on the vertebral column. Obviously, first the skin; then the muscles, bones, and ligaments; but never the spinal cord or its membranes, unless the bones or ligaments be destroyed. An inspection of the vertebral column will convince the reader at once of this truth. The length of the cervical spines, and the overlapping of the dorsal, not to mention the strong ligaments and massy muscles covering the transverse processes, render the spinal cord as secure from pressure from without as is the brain. Is spinal tenderness of any value as a diagnostic sign? The answer is, that disease of the vertebræ, and even of the cord itself, may go on to an extraordinary extent, with very little or no tenderness of the vertebral column, and with but slight functional derangement of the organs in connection with the spinal cord."

Nor must we forget, in considering this class of injuries,

* 'Nervous Diseases of Women,' p. 330.

that this widely-distributed "lumbago"—if the term may be allowed to indicate the aching and the pain on movement throughout every part of the spinal column—may give rise to a form of pseudo-paralysis which, if unrecognised, may cause unwarranted alarm to ourselves and to our patients. The pain in *all* movements may be so great, as shown in the second case, and seen in many others which have come under our notice,* whether the movements be of the limbs, of the body, or of the head, that the patient is really afraid to move at all. This well-grounded *fear of moving*† may soon assume the importance of an absolute inability to move. Ask any man who has had a severe lumbago, whether from a sprain, from rheumatism, or from cold, if he has not at the same time felt a strange sense of difficulty in moving his legs. Brisk walking becomes impossible; the effort needed to put one leg before the other must be unnaturally great; fatigue comes on early, and the patient complains to you that his legs feel weak and as if he could hardly move them. Free micturition may likewise be interfered with, from lack of the natural support and help which the lumbar muscles provide when this act is being performed. The patient perhaps cannot completely empty his bladder, and there is a certain amount of dribbling at the close of the act. It thus appears to himself that his "water runs from him," and if thereto, as a consequence of slight retention, there be added some irritability of bladder, symptoms of somewhat ominous import seem to be developed. This bladder

* See Cases 34, 88, 117, 124, Appendix, and many others.

† A case came under our observation not long ago in which the *fear of moving* was strangely manifested. A man who had received such injuries as we have described, and was confined to bed in consequence, needed three persons to help him out of bed every time he wanted to pass water in the day. To himself it appeared wholly unaccountable and extraordinary that whenever he woke in the night he could jump naturally out of bed without any help for the same purpose. It need hardly be said that the case was perfectly genuine.

trouble may rise into considerable prominence, especially when the nervous system has been much upset by the shock of the accident, and we may get a condition of "nervous bladder," in which the patient has a frequent desire to pass water with inability at the same time to perform the act perfectly, and consequent slight dribbling at its close. Constipation also arises from the same muscular incapacity, and becomes an almost invariable feature in the case. Thus it is nothing more nor less than natural for the friends to say that the patient is "paralysed," and paralysed from severe injury to the spine. If we do not avoid these fallacies, and do not correctly interpret this state of things, we shall add greatly to the dread, which after railway collisions may be very real, that "paralysis" is going to supervene.

In his very instructive and interesting book on 'Bone Setting,'* a book which will well repay the perusal of surgeons, Wharton Hood has the following passage on the pseudo-paralysis which sometimes complicates the sprain of a joint :—" A patient will unintentionally deceive his surgeon by saying that the affected joint 'feels weak,' an expression that seems naturally to suggest the use of some form of mechanical support. . . . . The meaning of 'weakness' in such cases is that the joint cannot be moved without pain, and people only use the word for want of knowing how to describe accurately the existing condition. Any one who has ever suffered from lumbago will understand this. A person so suffering feels 'weakness' in the sense that the power to rise from the recumbent posture is apparently gone. It is not really gone, but there is an instinctive dread of calling the affected muscles into action, and this dread conveys to the mind an impression of inability to move which can only be overcome by a most determined effort of the will." And precisely the same

* 'Bone-setting' (Macmillan, 1871), p. 29.

condition may be caused by the sprains of the whole vertebral column which we have considered.

In the absence of other signs of injury do not therefore let us give undue weight to this pain and tenderness at one or more points of the spinal column. When we meet with cases of real damage to the spinal cord or its membranes, we shall see of how small value this pain and tenderness become as signs of the disease. They may help to localise the point at which mischief is going on, but they do not indicate the mischief itself, nor are they in any sense pathognomonic symptoms of spinal cord disease. "I fear it must be admitted," Hood says,* "that the great importance of the spinal cord, and the gravity of its diseases, have rather tended to make professional men overlook the osseous and ligamentous case by which it is enclosed, and which is liable to all the maladies that befall bones and ligaments elsewhere." Of this we feel quite sure that were these conditions of vertebral column injury more frequently recognised and more correctly estimated we should hear less of the "railway spine" with its attendant evils, and we should see fewer errors in diagnosis as to the existence of disease of the spinal membranes or of the spinal cord. In no wise do we seek to lessen the real importance of these vertebral sprains. They may be exceedingly distressing to the patient; the pains may last for a very long time; there may even be occasional reminders of pain for months or years under suitable conditions; but it is right that we should attach no more import to them than they deserve, and that their existence should not entail a needless dread of serious injury to the structures within the spinal canal.

Had these pains which have been described been of the grave importance with which many so often regard them, a very different tale, indeed, would have to be told here of the injuries to the back to which collisions give rise. Met

* Op. cit., p. 144.

with so often, and with an almost unvarying persistency in so many of the cases which the surgeon has to treat after collisions, it is imperative to know what these pains are, and what they signify.*

Passing, then, from the pain and tenderness which are the more immediate signs of the injury which the back so frequently receives in collisions, it behoves us to speak of the other symptoms or complaints which occasionally accompany them, and which may be regarded as evidence of damage to structures other than the bony, the ligamentous, and the muscular, which are to be found in the vertebral column. We have spoken of the pseudo-paralysis which is due to an instinctive dread of inducing pain, whether by movement of head and neck, of the whole trunk, or any of the limbs; but occasionally we hear other complaints, of such sensations as tingling and numbness, or of "pins and needles," in some parts of the extremities.

If a man receives an injury, which has doubtless been severe, to his spinal column, and obviously to his spinal column alone, we should rightly regard any such complaints as worthy of our serious attention and anxiety, and more especially so if they lasted for any length of time. Injuries in severe collisions, however, are rarely limited to a single blow upon one part of the spine. It is more common for the patient to have been bruised and sprained in many and various parts of the body, and there thus arises another source of fallacy in the correct estimation† of the injuries received, and one which must be carefully excluded before we assume that the sensations complained of are the peripheral manifestations of some central damage. The proportion of cases in which these complaints are made is very small.

* Some evidence of their real gravity may be gathered from the appended table of cases, which will tell more than can be here written upon the subject.

† As in Cases 29, 68, 92, 109, 190, Appendix.

E. H. D.—, aged 35, received in a severe collision "a blow," as he expressed it, "down his whole back," and also on the back of his head from a falling carpet-bag. He did not consider himself much hurt, although from the account of his appearance there must have been a considerable degree of shock. He proceeded on his journey, but three quarters of an hour after the accident he felt compelled to stop and go to bed at a neighbouring inn. He then began to suffer from severe pain in the head, and from pain down the whole of the spine, but more especially about the sacrum and the lower cervical region. There were no marks of bruising. He also complained of "numbness and tingling" in his limbs, with some difficulty in moving them. He suffered for three days from extreme nervous prostration; dreaded the least noise; spoke only in a whisper, and lay in a darkened room. There was, however, no disturbance of pulse or temperature, and he had been able to sleep without narcotic for a few hours on the night after the accident. On the following days his limbs felt more natural, and the tingling and sensation of numbness had very much lessened. In five days these sensations had completely disappeared, but he still suffered from much pain about the vertebral column, and movements of the neck and trunk were painful to him. He was excessively nervous, and much dreaded any examination of his back. The pulse and temperature were throughout normal. He continued steadily to improve, and in three weeks was able to be moved. In three months he was going out daily, walking slowly about three miles a day, but complaining much—especially under examination—of pain in and about his vertebral column, the movements of which were evidently stiff and painful. He was still very nervous and felt generally weak, but there was no impairment of motion or of sensation in his limbs. He returned to work in about seven months. Five years after the accident he was at work and in good health, though often complaining of his back, and that "especially when lifting heavy weights."

It is obvious both from the history at the time and from the long-continued pain afterwards, that there was here precisely the same kind of injury, as far as the spinal column was affected, as in the two previous cases, but with the important addition of some abnormal sensations in the limbs, coming on synchronously with the pains, and disappearing after a few days. These sensations were general, and were not confined to any one limb or part of a limb; and although there is no evidence to tell us with certainty whereon the sensations depend, there are good grounds for believing them to be due to some effect produced by strain or blow upon the nervous cords proceeding from the spinal column to the limbs.*

In severe collisions, where there is a risk of the body being suddenly bent and strained in many different directions, it is highly probable from the cases cited, that every part of the spinal column is liable to muscular and ligamentous strain, and it is not inconceivable that the nervous cords which permeate the column at both sides should be involved in the same injury.

In this case, however, a very unusual kind of blow was received, a blow, as the man expressed it, "down the whole back;" and as we before observed, in recording Dr Bastian's very remarkable case of concussion lesion, the cord may under such circumstances be caught at a disadvantage, which its position and physical surroundings usually enable it to avoid.

In his classical work on 'Rest and Pain,' John Hilton refers to these peculiar sensations of "pins and needles." "What has happened," he asks, when they occur, for example, after a man has fallen with his back upon the ground? "It is possible (p. 49, ed. 2) that the spinal

* See also Cases 1, 44, 84, 89, 136, 188, Appendix.

marrow, obeying the law of gravitation, may, as the body falls, precipitate itself in the same direction, fall back towards the arches of the vertebræ, and be itself concussed in that way. Or the little filaments of the sensitive and motor nerves, which are delicately attached to the spinal marrow, may, for a moment, be put in a state of extreme tension, because, as they pass through the intervertebral foramina, they are fixed there by the dura mater; and if the spinal marrow be dragged from them, the intermediate parts must necessarily be put upon the stretch, producing at the time the 'pins and needles sensation,' and also explaining the symptoms felt on the following day."

At the same time we must be careful to remember that abnormal sensations in the periphery may undoubtedly be caused by some blow—altogether unsuspected—over the course of the nerves themselves. We have seen several cases of this kind where the point of definite injury has been found by careful examination, but where the sensations had been referred, with much alarm—in ignorance of the receipt of any blow upon the limb—to some damage to the contents of the spinal canal.

The abnormal sensations in such injuries are usually transient, not lasting longer than in the case just quoted, but they must not on that account be regarded as of no importance, whatever be their origin. They point out the advisability of absolute rest for some time after the accident, and the great risk that is involved in attempting to "walk off" the effects of the injury. "It is impossible,"* says Hilton, "that these symptoms could be the result of anything but some structural disturbance; and they are, to my mind, the evidence of decided injury to the nerves or marrow, although what that injury may be is not ascertainable. . . . The deterioration of function which follows such accidents (*i.e.* railway collisions) must be the effect of some kind of structural disturbance,

* Op. cit., p. 50.

for it occurs immediately after the blow. It is a shock to the spinal marrow. At least, that is the most reasonable light in which to consider it, especially with reference to the proper treatment.

"The object here should be to give the marrow rest from occupation by not allowing the patient to take walking exercise at all; or if at all, the exercise should be short of fatigue; certainly he should not be advised to endeavour to 'walk off' his condition. . . . The confirmation of the accuracy and applicability of these views is, I think, made apparent when it is added that all these morbid effects of concussion of the spinal marrow are to be prevented, relieved, or cured by due and long-continued rest."*

Happily the cases in which these sensations arise are very uncommon, and the symptoms themselves are usually transient, and we are therefore unable to say whether in any case they may be the precursors of less dubious evidence of central nerve lesion. They are usually found only after accidents of some severity, where there is a liability to receive considerable nervous shock with extensive bruising and straining of the whole body. Whatever be the condition on which these transient abnormal

* Abnormal sensations described as "tingling" or "pins and needles" are, although purely subjective, much more trustworthy terms, we think, than "numbness," which the laity use with a signification wholly different from that in common use amongst ourselves. Dread of moving the limbs without causing pain, bodily weakness and stiffness from confinement to bed, or the sensations produced by general bruising of the limbs, may each and all be described as "numbness," even when we find, and the patient admits, that there is no real anæsthesia nor true loss of muscular or sensory power. Only the other day we saw a man who described his leg as feeling "numb," the injury received having been a slight blow on the side of the knee. The use of this term is one which should be carefully inquired into and observed, for we may be altogether led astray if we fail to recognise the real sense which the patient intends to convey by it.

sensations depend, it seems to us very doubtful indeed whether they are due to any structural lesion, such as might be discovered if visual examination were obtainable during their continuance. They leave, as far as we know, no trace behind; and although the *precaution of absolute rest is imperative*, there is no evidence as yet to show that they are of more serious import than is the sensation of tingling in the ulnar distribution after a blow upon the elbow. It is not without significance, we think, in considering the cause of these sensations, and the possibility of their being due to some irritation of the nerves as they proceed from the spinal cord, or as they pass through the foramina of the vertebral column, to point out that, although we have been carefully on the look out for some such manifestation of central nerve disturbance, we have never seen a case in which Herpes Zoster has been met with after these injuries to the back. Some central nerve irritation is usually regarded as a cause of this cutaneous inflammation in the course of individual nerves, for it is often preceded by acute pain or some less obvious sign of nerve irritation or disturbance. That nothing of this kind should have been met with in these cases of severe general straining of the vertebral column seems to us not unworthy of remark, as affording negative evidence of the immunity of the nervous cords themselves from serious injury or disturbance, unless, indeed, the supposition be a wholly erroneous one that herpes has any central origin at all.* The existence of these abnormal sensations, how-

* A case is recorded by Dr Wilks ('Diseases of the Nervous System,' p. 243) of fractured spine and permanent paraplegia, where the patient " has often attacks of herpetic eruptions around the buttocks and backs of the thighs;" and Dr Buzzard has recorded cases where the " lightning pains " of tabes have been associated with recurrent attacks of herpes ('Brain,' vol. i, p. 168, *et seq*.). The most recent researches, however, seem to show that a peripheral neuritis is the frequent underlying cause of Herpes Zoster.

ever, does not militate against the broad conclusion that the spinal cord itself is very securely protected from injury in its osseous canal, and that we shall probably find more definite symptoms where there has been undoubted lesion of the marrow. There are, of course, degrees, from the most trifling to the most serious of lesions, no matter of what organ or structure, and it may perhaps be that these abnormal sensations are the symptoms of a disturbance which, in cases of more obvious local injury, amounts to actual damage—discoverable and giving rise to undoubted symptoms—of the spinal cord or of the nerves proceeding from it. We have said "obvious local injury," because we know of no case in which actual lesion of cord or nerve trunks has not been accompanied by local injury to the vertebral column, or in which the mode of accident has not been such as to afford presumptive evidence that the vertebral column itself has suffered damage. True, the difficulty is often very great of determining with certainty whether any blow has been received, for in the graver accidents, with their attendant crash and confusion, the fact that the patient has been dazed by the sudden shock disables him from giving any concise account of what happened to him at the time. Indications, however, are not usually wanting—by marks of bruising, by swelling, or persistent pain—that a severe blow has been received, and in the long run we shall find some guide to the degree of its severity by a knowledge of the nature of the accident in which the injury was sustained.

Before proceeding to give examples of undoubted nerve lesion received in collisions, let us again direct attention to this fact, which the record of a large number of injuries indisputably proves, that cases where there has been unquestionable lesion either of central or more peripheral parts of the nervous system are few and far between. Nothing stands out more clearly than this in the whole range of our inquiry as to the nature of the injuries to the back, which in the present day become the subject of medico-legal inquiry, that these lesions are met with only in a few isolated cases. It is essential that this fact should be duly recognised if we are to form a right estimate of the much wider class of cases where the injury and symptoms are rather those of general nervous shock, variable in degree, protracted in time, where it may be held by some that there has been concussion of the spinal cord or a "concussion of the spine," even though there has been no evidence of blow upon, or of damage to, the structure of any one organ or part of the body.

S. V—, æt. 48, the mother of several children, gave the following account of the injuries she received in a severe collision which happened at night, and in which a great many persons were hurt. She had no distinct remembrance of what happened to her, but after she had been carried home and put to bed, marks of severe bruising were found between the eyes, at the back of the head, on the chest, and more extensively about the lumbar and sacral regions of the vertebral column. Beyond the bruising there were no physical signs of injury to the spine. She was in bed for three months, during which time she suffered chiefly from pain in her back, so bad at first that she could not move in bed. She also had what she described as a "numbing pain down the left leg." At the end of three months she was so much better as to be able to get out, and shortly after to be moved to the seaside. Ten months after the accident her chief complaints were of pain in the back, especially on movement, and of a partial loss both of motion and sensation in the left leg. She was then using the leg as much as she could, though obliged to go about with crutches. There was *very slight* wasting of the limb, all the movements of which, it may be said, were sluggish and defective in power. There was diminution of ordinary sensation, slight only, but yet undoubted. There was no dragging of the limb in walking, the leg being rather held stiffly, and conveying the impression of injury having been received about the hip. There could, indeed, be no doubt that, in addition to the paresis, there was injury to the muscular and ligamentous structures about the hip and pelvis, injury which led to much of the stiffness about the thigh, such as might have been remedied by freer movement than the patient could give to it. There was never any paralysis of bladder or bowel. (See also Case 179, Appendix.)

It must, we think, be regarded as very highly probable that the impairment of motion and sensation in this case was due to some injury to the nerve trunks, but it is impossible to say with certainty whether that injury was after the nerves had formed the plexuses outside the vertebral column or when they remained as individual cords in the cauda equina. If the paresis was due to traumatic lesion of the nerve trunks within the spinal canal, it is almost inconceivable that the effects could have been confined to the nerves of one limb only, and on this ground it seems more reasonable to conclude that the injury to the nerve trunks was outside the vertebral column. The length of time, moreover, that elapsed before the recovery of the patient, seems further to confirm this opinion, whether the essential lesion was of the nerves themselves, or, as is equally probable from the character of the blow, was hæmorrhage lying around and pressing upon them.

From the time when this patient was seen she steadily improved. Thirteen months after the accident she was able to abandon her crutches, though the report of her then was that "sensation is still very feeble in the injured leg, but there is still progress made." Two years later, or three years after the accident, this last report was sent:—" The case you saw two years ago has resulted in a complete cure, and she is now able to walk about almost as well as ever." The patient was continuously under the observation and care of a most able practitioner, who took great interest in her case, and to whom we are indebted for this valuable knowledge about her.

It is not easy in a case such as this to separate the symptoms which are due to interrupted or damaged nerve function from those which depend on muscular, ligamentous, and osseous bruise and strain. These last may be very severe, and may give rise to great mechanical impairment of motor power, producing the pseudo-palsy to which earlier reference has been made; and only a very careful examination of the limb and of all the attendant symptoms will enable us to decide that there has or has not been injury to nerve structures. It is not surprising that, in cases like this, the view should often be taken that there has been "concussion of the spine with paralysis of one leg." How inappropriate is the term "concussion of the spine" to apply to them it is needless further to point out, and it would be better, we think, to call them what they are, cases of "severe bruising and strain of the muscles, ligaments, and articulations, and possibly of the nerves of the pelvis and thigh."*

We have spoken of the great importance of endeavouring to find out, as far as may be, the character of the blow which has been received in collision accidents, or the precise mode in which injury has been inflicted upon the back. The result of our own inquiries and experience leaves no doubt in our mind that when there is undoubted lesion of nerve centres or of nerve trunks, that lesion has been caused by localised injury at the part where the disease at first exists, and that general nervous shock is wholly inadequate to bring about so grave a result. The

* *Vide* a case recorded by Dr Buzzard ('Lancet,' vol. ii, 1879, p. 872) of "Paralysis with wasting of one lower extremity," due to a perineuritis of the nerves, but whether in the cavity of the pelvis or in the cauda equina it was impossible to say.

cases already given exemplify this, but not so markedly or with such terrible force as the following, happily the only case of its kind which has fallen under our own observation during many years:—

T. L—, æt. 29, a thin, delicate man, was in a very bad collision in which three persons were killed, and in which a large number were injured. He could give no clear account of the accident, saying he was dazed and could not stand. He was confined to bed for about a fortnight, suffering much from pain in his back and legs. He then improved a little, and was able to go to a hydropathic establishment in the country where he stayed for two months. His back continued to trouble him very much (this is in his own words), but otherwise he improved. Towards the close of this two months the weakness which he had all along felt in his legs became a much more definite loss of power, and in a week or ten days he was quite unable to walk. Ten months after the accident there was no mistake as to his condition. Loss of power to move, and almost entire loss of sensation in his legs, paralysis of bowel, paralysis of bladder with alkaline urine, bed sores, and reflex spasm of the lower limbs, were the undoubted symptoms of softening of the spinal cord. His pulse was frequent, and his temperature above normal. He complained of pain in the lower part of the back, *but there was no marked tenderness*. He lingered for some months without improvement and then died.

It is a misfortune that no post-mortem examination was allowed in this case, for it would have been in every way desirable to have learned exactly the nature of the injury, if any, which the vertebral column sustained, and how it came to pass that the cord became affected. The precise train of events can therefore be a matter of conjecture only. Two months elapsed between the receipt of the injury and the onset of definite symptoms of inflammation and softening of the spinal cord itself, and there is unfortunately no more accurate record of what his condition really was during this period.

The mode in which he was injured, however, seems to render it not improbable that there were symptoms which should have made a fortnight's stay in bed altogether inadequate, and so early a move to the country most unwise. For the accident really happened in this way, as was learned from another person who was with him at the time. The patient was thrown on his face on the floor of the carriage, and a very "heavy man then fell on the top of him, right in the middle of his back." There was here no doubtful history of a blow, of obscure injury to a remote part, or of a general shake to the whole body, but a clear account of an accident so happening that, without any visible signs of damage to the spine, there may well have been some separation of vertebræ, or some undue bending of the column which, damaging at the same time the membranes of the cord, or causing slight intra-spinal hæmorrhage, was the real starting-point of the mischief which supervened. This case exemplifies, better perhaps than any of which we know, the importance of learning and paying due regard to the precise history of the accident and of the injury, so that we may escape from the region of cloud-land where we hear no more than that a man has been in a collision, and had concussion of the spine and became paralysed.

Looking to the manner in which the back was in this case injured, it seems highly probable that the spinal column received a very severe wrench or sprain, and that in the sudden bend to which it was subjected by the falling of a heavy weight upon it, some separation may have occurred between two of the vertebræ. The important question is thereby suggested: How far does such an injury render the spinal meninges liable to inflammation, even though the spinal cord may itself have received no damage? We would suggest the following as a very probable sequence of events in the case recorded:—A small localised injury to the membranes, or hæmorrhage, at the site of the injury to the vertebral column, followed by a meningitis which was at first too limited in extent to give rise to any definite symptoms, but which preceded, and, having implicated the cord, was the cause of, the myelitis which had a fatal issue. If this be the explanation the meningitis must have at first been exceedingly limited, because *traumatic* meningitis does not usually follow an unobserved course. Sometimes it is acute, spreads rapidly, gives rise to very definite symptoms, and has a fatal result. At other times a less violent inflammation leads to local thickening and adhesion about the spinal roots causing peripheral symptoms of impaired innervation; or similar pathological changes may involve the cord itself, and setting up degeneration therein present very definite symptoms of structural disease. But acute myelitis at so long an interval after injury is most uncommon. Still further doubt, however, surrounds this case, for at the time when the symptoms of spinal softening were becoming marked the man had orchitis. It is true that gonorrhœa was altogether denied, and that the orchitis may have been caused by the frequent drawing off his water by the catheter; but it cannot be forgotten that myelitis of the cord has been proved to be the result of thrombosis of the pelvic and vesical veins, a direct consequence of the same urethral inflammation which had produced orchitis. It seems, therefore, within the range of possibility that the myelitis and the fatal result were not due to the injury at all.

We have not in our possession the notes of any case of acute traumatic meningitis following railway injury, nor have we seen a single case in which we could satisfy ourselves of the presence of "subacute" or chronic meningitis as the basis of the symptoms which are so frequently seen after the shock of railway collision, those symptoms of general nervous shock to which in a coming chapter we shall refer in detail. Over and over again we have heard "subacute meningitis" put forward as the cause of the localised pains in the back which are so common after sprains of the vertebral column, and this more especially when the pains have been associated with general bodily weakness due to long confinement in bed or in the house, and with the nervousness and emotional disturbance which are inseparable from severe shock to the system. The history and course of the cases, however, coupled with an entire absence of the real symptoms to which meningitis gives rise, have convinced us that no such serious mischief could exist. For meningitis of the spinal membranes is a serious mischief, and it is hard to believe that if subacute meningitis of traumatic origin be so common as some would seem to believe it to be after collisions, we should not much more frequently meet with cases of meningitis running an acute course, or with cases where there is likewise degeneration of the spinal cord. The burden of proof as to the existence of chronic and subacute meningitis in these very common cases of injury to the back lies, it seems to us, with those who would refer the usual symptoms to this pathological condition. The existence of this lesion should be established, it is only fair to ask, upon positive grounds, even though the difficulty of proving the negative may not in this instance be so great as is usually and proverbially supposed.

Far too commonly it is supposed that a local pain over one or more vertebræ—such pain as we meet with in cases of simple sprain—is an evidence of spinal meningitis. We shall do well, however, to note the following passage from Dr Gowers,* pointing out where we must look for distinctive evidence that the meninges are thickened and inflamed :—"The only symptoms," he says, "which are

* 'Diagnosis of Diseases of the Spinal Cord,' London, 1880, Churchill, p. 67.

usually due to this condition (meningitis) are those which result from the involvement of the nerve roots in their passage through the diseased membranes. The roots are irritated by the adjacent inflammation. The meninges often become thickened, and the change is then called 'pachy-meningitis.' In this thickening the nerve roots are often greatly damaged by pressure. The irritation affects first the sensory roots, causing 'excentric' pains and hyperæsthesia, to which are often added areas of anæsthesia here and there, due to the greater damage of some nerve roots. The affection of the motor roots causes symptoms similar to those of disease of the anterior cornua, but very irregular in distribution. The peripheral motor nerve fibres, cut off from their motor cells, degenerate, and the muscular fibres waste and present electrical reactions which vary according to the rapidity of the morbid process. Sometimes the nutrition of the skin suffers. There is often, in addition, pain in the back, from the lumbar to the cervical region, sometimes severe between the shoulders. . . . . It must be remembered that inflammation often affects the substance of the cord as well as the meninges, or the cord may be pressed upon by the thickened membranes, and so mixed symptoms may result."

And here we must again observe how frequently syphilis * lies at the root, as it were, of any actual disease

* See an important paper by Dr Buzzard on "Cases of Syphilitic Paraplegia," 'Lancet,' vol. i, 1879, p. 469. After detailing a case of paraplegia, with strangely variegated symptoms, cured by anti-syphilitic remedies, he writes: "With the clear history of syphilis before us, we can well imagine that the meningitis was gummatous, and that it probably involved especially the internal surface of the dura mater (pachymeningitis), but extended also to the contiguous soft membranes. . . . . Note, also, that there was no tenderness on percussing the vertebral spines. The absence of this symptom is too often regarded as being almost inconsistent with the existence of serious lesion of the cord or its coverings. Nothing can well be less founded in fact. If

of the membranes of the spinal cord. There was a case under our observation in hospital not long ago where injury was said to have given rise to various symptoms— slight wasting of the legs, partial loss of motor and sensory power in unequal degree and extent in the two limbs, due to scattered meningitis; but in reality the injury was wholly inadequate to produce such a result, and there was a very distinct history of syphilis, and the symptoms had more than once been lessened and almost removed by the administration of anti-syphilitic remedies. And although from a medico-legal point of view the presence of a syphilitic taint does not debar the sufferer from compensation for injuries received, such taint must of necessity be acknowledged if we are to form a correct estimate of the probabilities there may be of recovery in any particular case. Grounds for hope of amelioration in the symptoms there truly are when meningitis of the cord is due to syphilis;* but it cannot be too strongly insisted that a traumatic meningitis is a wholly different thing, and that the tendency of this to pass away under treatment is infinitely less. We make this remark more especially, because we have often heard it given in evidence in courts of law that an obscure injury has originated a state of chronic meningeal inflammation or thickening, and that the prognosis is one of recovery in a few months, usually from six to eighteen; whereas we should say that such a prognosis is altogether inconsistent

we put aside cases in which the vertebral column itself is diseased, we shall find that the existence of very marked spinal tenderness points strongly in the direction of a functional nervous affection of comparatively little importance, and does not indicate a serious organic lesion of the spinal cord."

* " Inasmuch," says Dr Ross ('Diseases of the Nervous System,' 2nd ed., vol. i, p. 293), " as there is no severe organic disease affecting the nervous system in which the results of treatment are so often satisfactory, so there is no disease which deservedly brings so much discredit upon the practitioner who overlooks its presence."

with clinical experience, and that the chances of recovery are very distant indeed. True, the prognosis of recovery is usually verified, often far sooner than in the time allowed. The result is happy, but it is destructive of the diagnosis which was made.

Of the exceeding rarity of spinal meningitis as an immediate result of localised injury to the vertebral column we are well assured. Much more, then, must the same remark apply to the results of the injuries which we have considered in the early part of this chapter; and we know of no one case either in our own experience, or in the experience of others, in which meningeal inflammation has been indisputably caused by injury to some part of the body remote from the vertebral column.

The cases which we have thus been able to record of unquestionable nerve lesion are happily very few and far between, and they stand out as individual, isolated exceptions from those· far more common injuries where there has been no structural damage to nerve centres. They are, each of them, examples of the possible results of injury to the back without apparent mechanical lesion; but while instructive in themselves, their very rarity should preserve them from becoming sources of perpetual alarm to those who have been in collisions. Let this also be noted of them that the symptoms of definite structural lesion supervened within a comparatively short time of the accidents, thus helping to establish no uncertain bond between the cause and the effect. Cases like those we have given are happily free—and it is perhaps the only bright spot about them—from the glamour of unreality and imposture; and in not one of them did litigation add to the suffering, already ample, of the patients themselves.

It is a very remarkable fact, we think, that the injuries of the spinal column which we have endeavoured to describe, should so seldom be followed by distinctive evidence of spinal caries. The influence of injury in originating caries of the bodies of the vertebræ is well recognised, and although it is not easy in many cases to connect the disease with the injury supposed to have given rise to it, there can be little doubt that angular curvature in children has frequently had origin in strain or wrench of the spinal column, whereby the intervertebral substances have been squeezed and bruised, or the bodies themselves have received more direct injury. It is singular how the prevailing views about "concussion of the spine" seem to dominate over and exclude every other explanation of the common spinal injuries of collisions, and that we rarely or never meet with cases where the persistent pain and tenderness, which we have said are due to vertebral sprain, are attributed to the existence of spinal caries induced by the accident. And yet it would appear that spinal wrenches are quite as likely to set up spinal caries as to set up mischief in the membranes of the spinal cord. The immediate injury is inflicted on the vertebral column, and only in the very extremest degrees of sprain and wrench, where there is reason to think that some vertebral dislocation or separation may have occurred, do we meet with the objective signs—and that, too, at a comparatively early date, according to the extent of the injury —of actual lesion of the contents of the spinal canal. Common enough though these sprains and wrenches are, we have as yet met with no case where spinal caries and ultimate curvature have been produced by injury to the back in a collision.

A very remarkable case came not long ago under our observation which, apart from the interest derived from itself, shows what difficult questions may sometimes call for consideration and decision in cases of medico-legal inquiry. A strong and healthy girl, æt. 20, was hurt in a slight collision, more frightened perhaps than hurt, and had at once the usual pain in the back. Examination revealed, within a very few days of the accident, a projection in the middle line of the lower dorsal vertebræ, a projection very like angular curvature, obvious when she was sitting up, but much less marked when lying down. The friends had no knowledge whatever of the girl having been ill in childhood, nor were they, or she herself, aware of any abnormality in the back. They naturally attributed the condition to the accident, and prudence demanded treatment accordingly. The opinion was, however, expressed that the projection could not have been caused by the accident, because the general symptoms of such an injury—if injury at all—must of necessity have been far more grave. It was thought to be congenital. Months passed away without alteration in the state of the back, and the parents came at length to the conclusion that they had been mistaken, and the original opinion was confirmed—to their great comfort—by the foremost British surgeon of the day.

In all the cases which we have thus far considered one symptom has been almost invariably present—*pain in the back*. It becomes of importance, therefore to inquire into the real value of pain in the back as a symptom of grave spinal disease.

It is a daily observation in surgical practice that local pain at the site of disease is not one of the most pronounced symptoms of spinal caries, and that it is of infinitely less value as a diagnostic aid than pains at the periphery, instinctive dread of leaving the recumbent posture, or an absence of natural flexibility in the spinal column when the patient moves. Pain "is a fallacious monitor in regard to disease of the spine. It fails to warn when danger is imminent, and it alarms needlessly."[*] The same remark applies with no less force to pain as an evidence of serious disease in the spinal cord. "Pain," writes Dr Gowers,[†] "referred to the spine, is occasionally

[*] 'Holmes's System of Surgery,' vol. iv, p. 106; Shaw's 'Essay on Diseases of the Spine.'   [†] Op. cit., p. 41.

present in organic disease of the cord, but is more frequent in disease originating in the meninges or bones. But the frequency with which spinal pain is present in abdominal, especially gastric, disease, and in neuralgic affections, lessens the diagnostic value *when it exists alone.** It is probably no exaggeration to say that of one hundred patients who complain of spinal pain, in ninety-nine there is no spinal disease." "In all spinal affections," Dr Wilks says, "we look to the back, to discover if there be any disease in the vertebral column, and we generally percuss it. Now, as regards any value to be derived from this method, I think we must set it down as very small. We, of course, examine the spine, for by so doing we may discover a projection or a growth; but as for informing us of the condition of the medulla within it, percussion seldom does that. Of course, should disease exist between any of the vertebræ, any violent jar on the back would be likely to produce discomfort; but, as a rule, in slowly progressing disease of the cord, as in the majority of cases of paraplegia which we meet with, there would be no pain produced. At the same time a sensitiveness of the spine is very common, but this generally implies a simple functional hyperæsthesia, so that I verily believe that were you to test the value of this method of diagnosis by the rule of averages, you would find pain mostly absent in organic diseases of the cord, and present in those persons who suffered merely from nervous excitability."†

We could quote no abler or more experienced authorities upon the subject. It must, however, be specially noted that the statement is not that there is no spinal pain in cases of vertebral disease or of organic disease of the cord, but rather that the great frequency of spinal pain in numerous affections where there is certainly no structural disease in the vertebræ or cord, renders it by itself a sign

* Italics our own. † Op. cit., p. 199.

of but small value of the presence of structural disease. But yet we have seen that in the very common and ordinary cases of injury of the back, which in the present day become the subject of medico-legal inquiry, pain in the spine is a prominent symptom, exciting the alarm of the patient, and leading too often to an erroneous diagnosis of the nature of the injury. What difference is there between the two kinds of cases? In the one class, where there is progressive organic disease, be it of vertebræ, meninges, or cord, spinal pain is usually the least prominent diagnostic symptom, so little prominent, indeed, that *in presence of other and more obvious signs* it is regarded as of only secondary moment, a * "subordinate symptom;" while in the other class of vastly more numerous cases, spinal pain is *the* prominent symptom—perhaps the only one present at all—indicative of some injury about the vertebral column. Take a case of fracture of the dorsal spine with complete paraplegia. Is pain in the back the important guide to a diagnosis of the injury? By no means. We seek for those symptoms which physiologically reveal the injury and point out the damage which has been done, symptoms which in other cases are absent, and leave the spinal pain to tell its own unaided tale. And when it exists alone, or whether it be associated with some spinal tenderness also, we feel sure that in these cases, met with after railway collision, it will be found that, with the rarest exception, it is indicative of nothing more serious than muscular or ligamentous strain.

The pain is almost characteristic of the injury. Of a constant wearing nature, both in and near the vertebral column, and in the region of the muscles on either side, the pain is liable to be rendered acute by movement, so that the patient may dread to move at all; or if he be able to move about, it is likely to seize him suddenly and

* 'Holmes's System,' op. cit., p. 107.

sharply, to make him cry out for the moment, and to be followed by renewed aching of the back. The spinal pains are, indeed, very like the pains which may affect any contused or sprained joint; and we have precisely the same indication for treatment, that, after a time, when all inflammatory mischief—if, indeed, there has been any inflammation—has subsided, movement, although perhaps giving rise to sharp pain at first, is really the best thing to get rid of or to lessen the suffering. We have seen this happen too often in the treatment of these spinal pains not to make us feel confident that when adequate rest has been given to recover from the immediate effects of the shock, and to avoid what risk there may be from too early movement of the bruised and strained parts, movement—and not fixation of the whole back in poroplastic jackets—is essentially the best treatment for relieving the spinal pain. True, the effort necessary to move for the first time may have to be very great, and the pain on first getting out of bed or attempting to walk may be so acute as to drive the patient back to bed again, yet if this initial pain will be endured, and a little perseverance and determination be exerted, it is certain that the pain will be daily lessened, and that greater freedom of movement will be at the same time gained. In a sprained joint we well know how stiffness and pain go frequently *pari passu* together, and the same associated phenomena in these injuries of the back form an additional aid to the diagnosis that these spinal pains are not of more serious import than those due to muscular and ligamentous strain elsewhere. They share this feature likewise in common with them, that they have a tendency to last obstinately for a long time; to recur after intervals of comparative or entire ease; to be induced by changes in the weather, or by any extra exertion; and so, by their very nature, they unduly alarm the sufferer, and encourage him in the belief that his injuries have been greater than they really are, and that the prospect of his recovery without permanent damage or disablement is very remote indeed. It behoves us not to share in his alarm, but to use our every influence to induce the patient to take those steps which alone can ensure his recovery—to leave his bed, to have change of air, if need be, to improve his general health, and to allow of greater opportunities of movement than were he to stay at home. Let us be careful, however, before all things, that we do not overlook any symptom which is a real indication of injury of the spinal membranes or of disease of the spinal cord.

Unhappily we are too often frustrated in our desires to induce the patient to adopt those means which alone are needful to bring about relief from pain, by his inability to make the necessary effort to resume his work until he has received compensation for his injuries. And thus it comes to pass that at the end of many months the patient is no better than he was soon after the accident—it may even be that his back is more stiff and painful than before—and an additional proof seems to be afforded of the gravity of his injury, and of the improbability of his recovery. The contents of his spinal canal are supposed to be involved, and a hopeless prognosis is, perchance, proclaimed. But it will invariably be found that when settlement of his claim allows the requisite effort to be made, the spinal pain very soon subsides under constant movement, and it is not an insuperable bar to his habitual work.

There is yet another kind of pain very often met with in these cases of injury which seems rather to be a mental offspring of the muscular and ligamentous pains affecting the vertebral column. We refer to the hyperæsthesia of the back, the development of which may occasionally be observed.

A young man was slightly shaken in a collision of no great severity, and in a few days he had the pains about the vertebral column such as have been described. He gradually recovered from the effects of the shake, but the aching in the back continued, and the spine was therefore more especially examined, not much attention having been paid to it hitherto. The examination revealed a point of tenderness on pressure over one of the dorsal vertebræ, at the point in fact where the sprain had probably been most severe. Within a day or two his back became so sensitive that he complained of, and shrank from, the very lightest touch of the finger on almost every part of it, whether over the spinal column or over the muscles at the side. He was so sensitive to touch that he endeavoured to avoid being touched at all, seemed even afraid to have his back looked at, and moved himself away with so much contortion as in itself to afford evidence of the absence of any serious mischief about the vertebral column or its contents. The hyperæsthesia was doubtless perfectly genuine; but in addition to the mode of onset, observe the inconsistencies of the hyperæsthesia itself. So great was it, that had it been real and not *imaginary*, it must have been unbearable for the man either to have rested against his couch or even to have borne the contact of his clothes. His hyperæsthetic back should have been protected under a glass shade, secure even from the pointed insults of a fly.

This, then, is the hyperæsthesia so often found superadded to the pain which is a real consequence of the vertebral sprain; and yet too frequently the inconsistencies thereof are ignored and the hyper-sensitiveness is regarded as another, and more telling, symptom of some inflammatory condition of the membranes of the spinal cord. It has little in common, however, with the hyperæsthesia or the "excentric" pains which are a result of irritation of the sensory nerve-roots, whether by thickening of membranes or otherwise; and it is unlike the zone, or girdle, of hyperæsthesia which may feel to the patient like a cord, or some other abnormal sensation at the periphery. The hyperæsthesia is too widespread over one area, and is at the same time too limited to the area which is the chief seat of attention. It is, moreover, unlikely that real irritation of the sensory nerve roots should give rise to hyperæsthesia upon the back alone. It is rather the natural outcome of that alarm which, both in hospital patients and in those more especially who have been in railway collisions, seems to be inseparable from injuries to the spine or back; and although undoubtedly a real condition to the patient himself, it is yet unreal and the product of his disordered imagination alone. (See also Case 108, Appendix, and numerous others.)

We need hardly dwell upon the pain and local tenderness of which there is no sign when the attention of the patient is otherwise engaged and directed from the affected part. These may be of the same nature as the hyperæsthesia, though much more often they have no real existence, and are heard from those persons only who are purposely exaggerating the effects of the injuries they received. Let it be remembered that a far more important sign than variableness of the pain under examination, is the very fact that the attention of the patient *can* be so easily diverted from the affected part. Call to mind a case of severe inflammation of a knee-joint and the examination thereof, and ask whether it is so easy—whether it is not well-nigh impossible—to divert the patient's attention from his knee when being examined, and whether he does not guard it with all the more conscious care because he is being asked questions wholly unconnected with the painful limb. The very readiness with which the attention can be removed from the tender back is the symptom of greater import than the spinal tenderness or pain.

In every medico-legal inquiry we have to deal largely with the prognosis and to consider the patient's future, and the direct consequences of his injury, whether remote or near. What risk is there, after the receipt of some obscure injury of the back such as we have been considering, that at a distant time there shall supervene, as a consequence of the injury, the symptoms of degeneration of the spinal cord? No more vital question can be asked in dealing with the history of railway injuries. We have very carefully gone into it, and we have endeavoured to meet with cases where there has been degeneration of the spinal cord as a remote consequence of spinal injury. Our inquiries have either been singularly unsuccessful— and they have been made by direct oral, and written communication with many professional brethren in all parts of the country—or we must admit that secondary and remote degeneration of the spinal cord, in cases where there has been no distinctive evidence of injury, is very rare indeed. We say *distinctive evidence*, for we hold that we cannot include amongst injuries to the spinal cord those "molecular disturbances" which must affect every tissue or organ in the body when subjected to any severe general shake or jar. Molecular disturbance is not necessarily molecular disintegration or pathological change, and there is no evidence to show that molecular disturbance is in itself a grave condition, or likely to have evil results, unless there should have been at the time some well-marked pathological lesion such as might *post-mortem* be discovered by the eye. Were "molecular disturbance" to be followed by pathological change as a direct result thereof, the consequences of unnumbered slight injuries would be serious indeed.*

* Several competent authorities consider that injury may be the cause of subsequent spinal cord degenerations. Dr Wilks, for instance, in the second edition of his 'Diseases of the Nervous System' (p. 293), says, in commenting upon the views here advanced, that he has long held the opinion that "a violent shaking in a railway carriage at the time of a collision will produce a stunning effect on the cerebro-spinal centres, and that this concussion may be the starting point of subse-

Have the victims of railway collisions, subjected as they must have been to "molecular disturbance" from the very nature of the accidents, have they—numbered by thousands since railway accidents began—afforded a larger proportion of those degenerative diseases of the spinal cord which have for years engaged the searching attention of our pathologists, than those members of the community who have not suffered the same influences? There is no evidence whatever that they have.

Take tabes dorsalis, for example, one of the commonest of degenerative diseases of the spinal cord. What influence has injury in originating this disease? We have ourselves met with no case where tabes has followed an obscure injury to the back received in a collision, but we have recently heard of a case in which it was sought to be shown that the disease had been caused by a spinal injury met with seven years before. In this case the symptoms were said to be those of undoubted tabes occurring in a man about forty years of age. Seven years before, he had been in a railway collision and had received a strain of the lumbar region which caused him pain for some time, but which had not prevented him from following his usual occupation. Seven years after the accident his general health began to fail him, he sought medical advice, and the nature of his malady was discovered. After so long an interval, and in the absence of any connecting link between the injury and the symptoms, it is hard to believe that the disease can have originated in the way supposed, and especially so when it is remembered that tabes is a disease affecting definite tracts of the spinal cord. If these had been really injured— and there seems no reason why they and they alone should have been picked out in the general "molecular disturbance"—there would surely have been distinctive evidence

quent morbid changes." Dr Bastian, Dr Sharkey, and others hold, we believe, much the same opinion with reference to injuries generally.

thereof before the lapse of seven years. The case must be taken for what it is worth. It shows well, we think, the difficulty there may be in arriving at a true knowledge of the cause of a well-recognised disease, and how very easy it is to create evidence of the remote traumatic origin of a disease like tabes. More interesting and important than this case, because the disease and its alleged cause were not separated by so long an interval, is that of a man aged sixty, who was in an accident of the most trivial kind when by some sudden jolting from application of the brakes while shunting he was thrown forwards and struck his forehead. He made no complaint at the time, there was no mark of bruise, and he had no notion he had been hurt. Ten weeks afterwards he began to have to strain in micturition, and as a consequence of this there came some hæmorrhoidal protrusion for which he sought advice. The case was then regarded as an ordinary one of piles, with probably some prostatic enlargement, and the man abandoned all notion of making any claim against the railway company, believing, as he was advised, that his illness was in no way due to any injuries received. The bladder trouble passed away, but in a few weeks more he began to have severe pains in his limbs, his gait became tottery, and his general health rapidly failed. More careful investigation then revealed absence of knee-jerk, and the Argyll-Robertson pupil, symptoms which showed that he was really the subject of tabes dorsalis, and threw light upon the bladder trouble which had been previously obscure. The diagnosis thus established, a claim was forthwith made on account of the disease having been caused by the accident. We believe, however, that there was no connection whatever between the two. The man had a slight blow on his head, and not until ten weeks had gone by did the first symptom of his disease—difficulty in micturition—declare itself. From that time onward the disease made rapid strides, but there is no evidence to show that his spine or spinal cord met with the least injury, and anything like a collision had not occurred. Nay, we believe that the relation of cause and effect between the accident and the disease would never have entered any one's head were it not for the prevalent notion that a railway accident—be it of whatever kind it may —produces, by concussion, some effect upon the cord. These cases are doubtless spoken of in the neighbourhoods where the patients live as cases of paralysis caused by railway accident from which they had never recovered.

The same lessons seem to us to be taught by the very able essay of Dr L. H. Petit, "De l'ataxie locomotrice dans ses rapports avec le traumatisme,"* which is the most exhaustive inquiry yet made into the traumatic origin of tabes dorsalis. It is impossible, we think, to read any of the cases which have been brought together by M. Petit without feeling that the evidence is somewhat unsatisfactory, and is deficient as conclusive proof that the disease did in any case recited have undoubted origin in the previous injury sustained. The paper is worthy of reference, though here of course only with much brevity. It is divided into several sections, but the only one which relates to our present inquiry is that on "Traumatismes antérieurs à l'ataxie;" for the remaining sections, "Traumatismes survenant dans le cours de l'ataxie;" "De l'influence des affections intercurrentes sur la marche de l'ataxie;" and "Influence de l'ataxie sur la marche des lésions traumatiques," although of great interest, do not specially concern us. The author refers to the opinions of several authors, Horn, Steinthal, and Topinard, as to the apparent origin of tabes in injury, and writes: "En 1866, parut le livre remarquable d'Erichsen sur les maladies du système nerveux consécutives aux accidents de chemin de fer. L'auteur étudie spécialement la commotion de la moelle causée par des chutes sur le dos, sur les pieds, la

* 'Revue Mensuelle de Médecine et de Chirurgie,' tome iii, 1879, p. 209, *et seq.*

tête, et la meningo-myélite qui en est la conséquence. Nous y trouvons des examples de myélite chronique avec phénomènes ataxiques, mais pas d'ataxie pure, déterminée par la sclérose des cordons postérieurs de la moelle. Du reste, le mot d'*ataxie* n'est pas mentionné dans le cours de l'ouvrage." M. Petit collects a series of cases in which there is a history of precedent injury, but which are not, in our judgment, adequate to establish the proposition that true tabes has even rarely a traumatic origin. It is impossible to go minutely through the evidence which M. Petit brings forward. It must suffice to quote two cases which may be regarded as typical of the whole series, in that none are more precise in their history, or more conclusive as to the real connection of the tabes with the injury which undoubtedly *preceded* the recognised onset of the symptoms. His first case is as follows:—" Obs. I (W. Horn, in Steinthal, Journal de Hufeland, ' Beiträge zur Geschichte und Pathologie des Tabes Dorsalis,' p. 24). Homme de trente-sept ans. Chute de cheval; reste plusieurs heures sans connaissance. Pas de symptômes morbides à la suite. Mais, l'hiver suivant, accès de douleurs et de crampes dans les membres inférieurs; puis engourdissement dans les jambes; démarche incertaine, chancelante; dysurie avec incontinence. Horn diagnostiqua un *tabes dorsalis* traumatique. Saignée et strychnine sans résultat."

And to take another case where the injury is supposed to have been inflicted on a part remote from the back:— " Obs. VI. Leyden, ' *Die graue Degenerationen der hinteren Rückenmarkstränge.*' Berlin, 1863, p. 260, obs. 5.

Mécanicien de quarante-cinq ans. Ni syphilis, ni rheumatisme; sueurs habituelles des pieds. Avant Noël, 1859, il lui tombe sur le pied gauche une barre de fer qui blesse trois orteils. Il plonge son pied dans l'eau froide, puis applique dessus de l'eau glacée. Suppression de la sueur de ce côté, et de l'autre un mois après. C'est de ce moment que date sa maladie. En février, 1860, douleurs lancinantes dans le pied et la jambe gauches, principalement dans le mollet et les adducteurs. Peu après, mêmes douleurs du côté droit, mais toujours plus faibles, quelquefois assez violentes pour provoquer de l'insomnie. L'ataxie suit la marche."

We do not for one moment deny or seek to extenuate the possibility of true tabes dorsalis being directly caused by injury—by a meningitis (*e.g.*) or a myelitis accidentally attacking *the exact site where the lesion of tabes lies**
—but it is impossible to concede to cases such as those we have quoted, that they have in any sense established the fact. We cannot doubt that the publication of M. Petit's valuable paper will be the means of directing special attention to the subject, and of leading us to examine, with closer scrutiny than heretofore, the evidence which may be forthcoming in future cases as to the origin of the disease. The inquiry is fraught with difficulty from this very fact, that being so insidious in its origin, and being shown perhaps for years by symptoms which to the patient appear of no more alarming import than a few occasional "rheumatic" pains, it is well nigh impossible to say when

---

* Dr Gowers, however, has suggested that a wide-spread myelitis may perhaps clear up and leave a degeneration of the cord in the posterior columns alone. He writes: "Degeneration apparently occurs, and spreads, in the posterior columns with great readiness; and it is common for an acute change, which extends more widely through the thickness of the cord, to clear from all except the posterior columns, and persist in those. This occurs in both local primary myelitis, or damage by the pressure of an adneural disease" (Gowers, "Syphilitic Neuroses," 'British Medical Journal,' vol. i, 1879, p. 304). It is obvious that if this be true, and if it be equally true that meningo-myelitis is a common result of "concussion of the spine" from the vibratory thrill or jar, tabes dorsalis ought to be a very frequent consequence of collision injury. That it is not is another fact against the meningo-myelitis theory. May this, however, explain Mr Gore's case?

the disease may have had its real beginning, and whether it may not have existed long before the receipt of the injury broke down the patient's health, and so allowed the symptoms of tabes to become more obvious and more rapidly developed.

The origin of this disease is clouded with obscurity, and although a history of syphilis is often to be obtained from its victims, and many think that syphilis is its cause, there are powerful arguments in favour of prolonged functional excitement and exhaustion of the sensory regions of the cord having a prominent place in its ætiology, a question which has been admirably dealt with by Dr Donkin in 'Brain,' vol. v, p. 443. We ourselves are inclined to believe, from careful observation of all his surroundings, that in the case we have recorded on page 150, the cause of the disease lay in his own home and was at work long before the accident.

As an aggravating cause there is every reason to believe that injury—not necessarily to the back—may tend to develop and bring to light the more striking symptoms of this disease, just as an exhausting illness may render the patient less able to cope with his malady, and so may indirectly reveal, and appear to him to be the cause of, the tabes whose existence had not hitherto been known.

Injuries to the back without apparent mechanical lesion may forsooth be the starting point of the disease—we do not for a moment say they are not—but the evidence, we believe, has yet to be adduced which shall establish it as a conclusive fact. The evidence, indeed, which he brings forward is regarded by Dr Petit himself as hardly sufficient to warrant any precise conclusions, and he is hopeful rather of directing attention to the subject, and of pointing out the lines upon which future inquiry may profitably run. Well-recorded and carefully observed facts are yet needed to establish the point, but in the meantime it behoves us, in the face of the existing uncertainty, to examine with all the more care and precision every case of injury to the back without apparent mechanical lesion, that in a medico-legal point of view there may be no injustice done to the injured person, should there by any sign or symptom whatsoever be distinctive proof that some injury has been inflicted on the contents of the spinal canal.

It is impossible to deal adequately with injuries of the back in their medico-legal aspects if we omit to devote some portion of this chapter to those cases of neurotic disturbances, which are often supposed to be due to a traumatic disorganisation of the great nerve centres, and which, associated as they very often are with those pains and points of tenderness in the vertebral column which we have previously and especially pointed out, are very liable indeed to have their symptoms referred to the important contents of the spinal canal. Here step in to perplex the clinical inquirer those numerous cases of functional, emotional, or hysterical disturbance which are so often met with after injuries where there has been—as in railway collisions there must naturally be—a great deal of alarm and mental shock. Medical literature contains to our knowledge no description of a class of cases such as those which have been engendered by the collision accidents of modern times, and whose symptoms, owing to the prevalent views about "concussion of the spine," are very frequently regarded as due to injury of the spinal membranes or the spinal cord. We refer more especially to those cases of severe general nervous shock where there has been no history of injury to any one organ of the body, no evidence of blow having been inflicted upon any single part, and yet, nevertheless, a condition of very serious illness supervenes. The symptoms to which we shall more especially refer in the chapters on general nervous shock, consist more essentially in an extreme degree of prostration, in which every function of the body and mind seems to be for a time enfeebled. They may come on very soon after the accident, or may be delayed for many hours; they are not uncommon in those who have received no bodily injury at all, and, as we shall have to point out, they are the continued expression of severe nervous exhaustion. Picture a case of this kind, and add to it sprain of the muscles and ligaments* of the vertebral column, with the resultant pain on movement, obstinate in its duration, and the tenderness—sometimes acute—on touch, and it is neither more nor less than natural—though a natural mistake—that the whole condition should be attributed to some alteration or morbid change of the " spine " which has been " concussed."

The jar or vibration of the collision must doubtless have shaken the spinal cord, as it has shaken the brain and every other organ also; but there is scant reason, it seems to us, for finding an explanation of the symptoms in degenerative changes of the important and vital structures lying within the spinal canal. Alarming though cases of this kind may be—and early death will show how truly grave they are in exceptional instances—their tendency is to get perfectly well after a longer or a shorter time. And inconsistent though it may at first sight appear that anything apart from the injury and the bodily condition itself should influence convalescence, it is none the less a

* "Sprains of the ligaments of the vertebræ, rupture of the complicated aponeuroses and muscles of the back, are common and enduring lesions, too obvious in their symptoms to need detailed description. They are, without exception, the most frequent cause of the phenomena assumed to be those following concussion of the spinal cord. They give rise to much local pain, to a rigidity of the spine, a difficulty in rising from the seat, a stiffness in walking, and contribute readily to any disposition on the patient's part to make much of his injury. The attitude, or the cautious and constrained movements of the body, may be made to suggest inferences which cannot be too guardedly accepted." (Dr R. M. Hodges on " So-called Concussion of the Spinal Cord," ' Boston Medical and Surgical Journal,' April 21st, 1881, vol. civ, p. 363.)

fact, which daily experience confirms, that the period of recovery very frequently begins at the moment when all mental anxiety and worry are at an end as to the legal aspects of the case. The settlement of the patient's claim for compensation has a potent influence in bringing about convalescence, not necessarily because there has been imposture or a lack of perfect genuineness in the facts and features of the case, but because, *as a very symptom of the malady itself*, there must have been an inability to bear the strain which a medico-legal inquiry demands. The strain removed, the anxiety lessened, there is nothing to stand in the way of a hopeful effort being made to return to a more natural and healthful mode of life, and each returning day of improved mental tone forges another link in the chain of progress towards recovery. We do not deal now, be it remembered, with those cases which are not strictly genuine, and where there is grave reason for suspecting the *bona fides* of the patient. Examples of this kind are unhappily too common, where the temptation to keep up the invalid state is great in expectation of larger pecuniary advantage to be derived thereby; but we must be careful that the absence of objective signs of nerve lesion does not lead us into the error of throwing doubt on the integrity and trustworthiness of those who are really suffering from the symptoms of general nervous shock. There is a vast difference, however, between the false and the true, and the impostor is usually not long in revealing himself by his actions, his symptoms, and his words.

CHAPTER IV

SHOCK TO THE NERVOUS SYSTEM

HAVING endeavoured thus far to give account of the real facts, as they seem to us, in connection with injuries of the back, concussion of the spinal cord, and concussion of the spine as seen after railway collisions, our next duty is to turn to the cases vastly more numerous, and we think even more important, to which under the name "general nervous shock" we have in our previous chapters referred. We use "shock to the nervous system" as a term applicable rather to the whole clinical circumstances of the case than to any one symptom which may be presented by the injured person. It is a phrase which in its very lack of precision appears suitable to describe the class of cases which we must now consider; for we shall see that the course, history, and general symptoms indicate some functional disturbance of the whole nervous balance or tone rather than structural damage to any organ of the body.

We are all familiar with the term "shock" as synonymous with the "collapse" which is a concomitant of all profound and sudden injuries, whether inflicted upon the head or upon some other part of the body. And this collapse or shock we are wont to regard as the immediate expression of lessened or annihilated function of the great nerve centres which preside over the vascular system, paresis of the heart and of the peripheral parts of the circulation being the essential factor in inducing the pallor and coldness which affect the whole surface of the body, and the mental enfeeblement which is due to impaired flow of blood within the brain.

It is not our purpose here to enter into any lengthened description of the nature of true shock or collapse. No matter how the injury may have been inflicted, provided only it has been sudden and severe, whether by railway accident or by the more ordinary casualties of every-day life which bring patients to our hospitals, shock or collapse in greater or lesser degree is invariably recognised as one of the features of the patient's general condition immediately after the injury has been received. The collapse may be lasting and profound, or it may be slight and speedily pass away, but in every case it is an immediate consequence of the injury, which, by its very suddenness and severity, has induced the paresis which primarily affects the great central organ of the circulation. Nor shall we occupy time by describing the history and symptoms of cases of shock or collapse with which all are doubtless familiar in hospital practice. There is, in fact, no condition which is more obvious, more striking, or more appalling than that of the seeming lifelessness which is an indication that some severe impression has been made by injury upon the nervous system. We have learned to look for it as an almost invariable concomitant in some degree of all the more serious accidents which are admitted into our hospitals; and one inquiry which the surgeon always makes about the cases of injury or accident admitted under his care is this:—Is there much shock?

We have used the term "lifelessness" to describe this condition of collapse, and, indeed, it seems a highly appropriate one to give to the state of shock from injury in its more serious degrees. There is a lowering of the vitality of every organ and function of the body, from mental activity and capacity to the least important function in the animal economy. And that which lies at the very foundation of most of the symptoms of shock or collapse is temporary paresis of the heart, and of the whole circulatory system. The slow, feeble, or almost annihilated pulse, the pallor of the lips and coldness of the extremities, the mental hebetude, the anæsthesia of the surface, the relaxation of the sphincters, the lessened secretion of the urine, the impaired muscular action, each and all are dependent in varying degrees on the paresis of the heart and vascular system, and on the impression upon the whole nervous system of which that is the first and most immediate result. "I do not hesitate to say," writes Mr Furneaux Jordan, in his valuable Hastings Essay on 'Shock,' "that in every case of shock there is at first, and for a longer or shorter time, diminished frequency of the heart's action."* And again: "The degree of enfeeblement of the action of the heart will serve as a fair index of the reduction of vitality generally."†

It is impossible to enter here into the pathology of shock. Much has yet to be done in elucidation of the subject, and especially by experiments on animals. The most recent, complete in itself, and very able account of shock in all its bearings is from the pen of Mr C. W. Mansell-Moullin in 'The International Encyclopædia of Surgery' (vol. i, p. 369, Macmillan, 1882). He sums up the results of experimental physiology by saying that, "In short, shock is an example of reflex paralysis in the strictest and narrowest sense of the term—a reflex inhibition, probably in the majority of cases general, affecting all the functions of the nervous system, and not limited to the heart and vessels only." M. Duret, however, regards the phenomena of concussion as due to changes in the tension of the cerebro-spinal fluid rather than to any effect upon the cerebro-spinal mass itself.

Collapse, as we well know, may be so profound that it

* 'Surgical Enquiries,' 2nd edit., p. 12.   † Ibid., p. 22.

stands out pre-eminently as the chief source of danger to the patient, a greater danger, it may be, than the bodily injury which he has sustained. A fatal issue may rapidly ensue, but happily death is the exceptional rather than the usual result of uncomplicated shock or collapse. The heart under appropriate treatment and care regains its normal power, and the functions are once more naturally performed. The collapse may be early succeeded by a period of "reaction," in which the temperature and pulse are slightly raised; but whether there be reaction or no, the symptoms of collapse pass away, and except as an indication for treatment, in our hospital practice, as a rule, they give us little more concern. Patient and surgeon are far more interested, after the subsidence of the initial collapse and its incidental risks, in the successful issue of the operation it may be, or in the usefulness restored to the limb or organ which has been damaged. If these results be satisfactory, the patient leaves the hospital gratified at the recovery which he has made, and thankful perhaps that his injuries were not more severe. What has become of the collapse which on his admission into hospital was a startling feature in his case? Has the shock no after-symptoms, or has it, as we thought it had, completely passed away? Important questions these; but as a matter of fact we know little of the after-history of our hospital patients, and from their cases we can draw but little help in elucidation of the general nervous shock which we meet with after railway collisions, and whose after-symptoms may perhaps be far more prominent than those which immediately followed the injury. Our hospital patients as we believe recover, and rarely or never do we hear of the symptoms, nor do we meet with the class of cases, to which we wish to direct attention now.

Surgeons are pretty well agreed, we think, that the collapse in cases of accident brought into hospital is usually very profound after the injuries—crush of limbs, for example—which railway servants meet with at their work, and which so frequently call for operative interference and aid. Compare two cases of like injury, the one received by accident on a railway, and the other by being knocked down and run over in the street, and the probabilities are great that the manifestations of shock will in the former case be more extreme than in the latter. And the difference lies in this, that in the one case there is an element of great fear and alarm, which has perhaps been altogether absent from what may be called the less formidable and less terrible mode of accident. How largely fright may of itself conduce to the condition recognised as shock is well shown by a case communicated to us by a surgeon of large experience, who, summoned to a railway station to see and conduct to the hospital a railway servant who had had his foot, as was supposed, run over on the line, found him in a state of collapse, and in greatest alarm as to the injury to his limb. Upon examination it was discovered that the only damage was the dexterous removal of the heel of his boot by the wheel of a passing engine. And medical literature abounds with cases where the gravest disturbances of function, and even death or the annihilation of function, have been produced by fright and by fright alone.

It is this same element of fear which in railway collisions has so large a share—in many cases the only share—in inducing immediate collapse, and in giving rise to those after-symptoms which may be almost as serious as, and are certainly far more troublesome than, those which we meet with shortly after the accident has occurred. The reasons for this are not far to seek. The incidents indeed of almost every railway collision are quite sufficient—even if no bodily injury be inflicted—to produce a very serious effect upon the mind, and to be the means of bringing about a state of collapse from fright, and from fright only. The suddenness of the accident, which comes without warning, or with a warning which only reveals the utter helplessness of the traveller, the loud noise, the hopeless confusion, the cries of those who are injured; these in themselves, and more especially if they occur at night or in the dark, are surely adequate to produce a profound impression upon the nervous system; and, even if they cause no marked shock or collapse at the time, to induce a series of nervous disturbances at no distant date. "The principal feature in railway injuries," says Mr Furneaux Jordan, "is the combination of the psychical and corporeal elements in the causation of shock, in such a manner that the former or psychical element is always present in its most intense and violent form. The incidents of a railway accident contribute to form a combination of the most terrible circumstances which it is possible for the mind to conceive. The vastness of the destructive forces, the magnitude of the results, the imminent danger to the lives of numbers of human beings, and the hopelessness of escape from the danger, give rise to emotions which in themselves are quite sufficient to produce shock, or even death itself. . . . . All that the most powerful impression on the nervous system can effect, is effected in a railway accident, and this quite irrespectively of the extent or importance of the bodily injury."*

In these purely psychical causes lies, we believe, the explanation of the very remarkable fact that after railway collisions the symptoms of general nervous shock are so common, and so often severe, in those who have received no bodily injury, or who have presented little sign of collapse at the time of the accident. The collapse from severe bodily injury is coincident with the injury itself, or with the immediate results of it, but when the shock is produced by purely mental causes the manifestations thereof may be delayed. Warded off in the first place

* Op. cit., 2nd edit., p. 37.

by the excitement of the scene, the shock is gathering, in the very delay itself, new force from the fact that the sources of alarm are continuous, and for the time all prevalent in the patient's mind. "In certain temperaments, wrought into a state of extreme excitement, a comparatively severe injury may not be attended with that degree of shock which, under other circumstances, would be seen. In those cases, however, shock is usually deferred, and not altogether averted; and it may be all the more severe, seeing that reactionary mental exhaustion, itself a kind of shock, is superadded to the effects of bodily injury."*

"We often see that the stimulus of Fear prevents fainting for just so long as it operates, and that directly it is withdrawn, the system yields to a reaction. Persons perform deeds of heroism in the immediate presence of danger, who do the right thing after the danger is over—swoon away."†

Due weight must therefore be given to alarm as a cause of the symptoms of nervous shock so frequently seen after railway collisions. We are inclined to think that sufficient importance has not hitherto been attached to it, and that many errors in diagnosis have been made because fright has not been considered of itself sufficient—as undoubtedly it is sufficient—to bring about the train of symptoms which we shall seek to describe. On the one hand, we may hear the symptoms regarded as evidence of serious and irremediable pathological change in the chief centres of the nervous system; and on the other hand, no clear history of pronounced shock or collapse at the time of the injury being forthcoming, the symptoms are deemed unreal and the *bona fides* of the patient is called in question. The mistakes are at opposite ends, and we know not which is the worse for the patient, who, really suffering and ill, lies

* Furneaux Jordan, op. cit., p. 27.
† Hack Tuke, 'Influence of the Mind upon the Body,' p. 249.

in the condition in which we find him because his whole nervous system has received a shock, not, it may be, from severe bodily injury which shows itself in unmistakable signs, but from the impalpable element of alarm, which must be measured by the events of the accident itself, and by the temperament of the individual who has been affected thereby.

The indications of collapse at the time of accident are very variable. The profoundest grades are occasionally seen after accidents of the greatest severity, where there has been in all probability destruction of life and limb. In these circumstances it is no marvel that examples of most alarming collapse should be met with, associated with definite structural injury, such an injury as would in fact be commonly marked by collapse, however and wherever it might have been received. It is a seemingly anomalous and most noteworthy fact that the collapse which in these railway accidents accompanies serious bodily injury, such as severe laceration of limb or fracture of bones—always excepting the collapse from severe concussion of the brain—is not followed, or is indeed very rarely followed, by the train of after-symptoms indicative of general nervous shock. This is a fact of the greatest interest and importance, and one which will help to throw light upon those symptoms of general nervous shock which are so often seen after the slighter degrees of initial collapse.

More numerous than the cases of profound collapse are those where the accident has been less severe in its effects upon life and limb, and where the earliest signs of shock have been comparatively slight. "I was thrown forwards and backwards in the carriage; I felt myself shaken but did not think I had been much hurt; I got out of the carriage and was able to help some of the other passengers, and I came on home by the next train:"—such is perhaps the simple story of the man who finds himself in a few hours, it may be only after two or three days, compelled to take to his bed because he feels so unnerved, and shaken, and ill. You make inquiry as to the more immediate effects of the accident upon him, and you perhaps learn that he felt shaken and was obliged to have some brandy, that he felt sick and faint for a few moments, or that he even vomited. He thought little of it, however, and gave help to those who needed it. A few hours elapse and he finds he cannot sleep; he has aches and pains in various parts of the body, most likely in the back; he feels as if he had been beaten all over; he is thirsty, feverish, and ill; and gathering fresh alarm from the very fact that he thought he had happily escaped all injury, he sends for his doctor, who sees that the symptoms of nervous disturbance and prostration have already begun. With varied modifications in detail, this is the history so often heard of the effects which an accident has had upon numbers of patients. It is clear, we think, from what happened at the time and from the early symptoms of reaction afterwards, that there was undoubted shock immediately after, or within a short time of, the accident. Slight it was in all its early manifestations, so slight perhaps as to call for little or no relief, disregarded as little more than a feeling of faintness or of being dazed, but enough to show that the alarm of the accident produced an instantaneous or early result, and to be the starting-point of that disturbance of the nervous system which may assume an aggravated form, and continue for a very long time. It will be part of our duty to explain how it is that the after-results of slightest initial shock from railway accident should be so often more lasting and serious than are the later nervous symptoms of those in whom the early collapse with extensive bodily injury has been more profound.

Lastly, at the other end of the scale we find the cases where there is no history whatever of injury or of the symptoms of collapse, no faintness, nausea, or vomiting, no early reaction from an initial stage of depression, but where the after-history very closely simulates that of the more numerous cases which fall under our care. These we shall consider by-and-by; for few are commoner than cases of spurious nervous shock.

Let us now illustrate by examples the kinds of symptoms with which we have to deal, and begin with a case of some severity, where there was undoubted collapse from the bodily injury received and from the very distressing surroundings of the accident; collapse both from bodily and mental causes. S. W—, æt. 46, a tall, somewhat powerful man, was in a very severe and destructive collision. He received bruises over both arms and legs, and also a blow upon the face which abraded the skin over, and fractured the bones of, his nose. He was not stunned. He lay for several days after the accident in a state of great nervous depression, with feeble and rapid pulse, and inability to eat or sleep. He suffered at the same time much distress from the fact that a friend sitting beside him in the carriage had been killed; and this seemed to prey constantly upon his mind. The bodily injuries progressed rapidly towards recovery, and in seventeen days after the accident he was able to be moved home. Nine weeks after the accident he had fairly well recovered from his injuries, and made no complaint of bodily sufferings. The ordinary functions of the body were natural, but his mental condition showed extreme emotional disturbance. He complained that he had suffered continuously from depression of spirits, as if some great trouble were impending. He feels very upset at our visit and begins to cry. He says he used to cry whenever he spoke to any one, but that now he has rather more control. He has been out of doors for a few yards, but was stopped by a sudden sensation as if his breathing were very short. His voice is very weak and indistinct, and occasionally he says it is almost inaudible. There is no disease of the larynx or adjoining parts. He sleeps very badly, waking frequently, and being constantly troubled by distressing dreams. His pulse is weak, 104. He occupies himself by a little reading and by occasionally going out, but he feels so shaken and weak that he is unable to do anything more. In many respects, however, he is improving. The weight he lost is being regained. He can walk rather further, is not so prone to cry, and his voice is stronger. Examination discovered no structural disease, but he was evidently in a most depressed and feeble state. Words, in fact, fail adequately to pourtray the distressing picture which this otherwise strong and healthy man presented. He remained in much the same condition for several months, though with undoubted tendency toward improvement. Fifteen months after the accident, several months, that is, after his claim had been settled, we learned that he was better, though yet very far from right, and he was considered wholly unfit for work. His history, given four years after the accident by his medical attendant, is as follows :—" In my opinion he will never be anything like the same man again. His appearance is much altered. He looks much older, haggard, and has become very bald. His voice is very weak, almost gone at times. For some time he went about in search of health, but improved very slowly, if at all. Lately he has obtained two posts, the work at which is of a very light nature. I just jotted down the following symptoms as he mentioned them, and I feel sure he would not wilfully exaggerate them. Very depressed spirits, sometimes palpitation, loss of sleep, bad dreams, very easily tired, can't walk more than two miles, then gets very tired and quite loses his voice. Did nothing for two years after the accident. Has lost all his energy. Sometimes has a great dread of impending evil. He *can* travel by railway without feeling nervous, but can't drive without feeling frightened all the time. I may add that his heart sounds are rather feeble, but not otherwise abnormal. Pulse 72. No special spinal symptoms; no paralysis; no bladder symptoms; always gets much upset if dining in company or if many people are talking near him. I knew him well before the accident and he was a very energetic and very honorable man."

It must be pretty obvious, we think, from his history, that this man's prolonged illness has been due in only small measure to the *bodily* injuries which he received. From these injuries, indeed, he had soon recovered, as soon, in fact, as if they had been inflicted in any other accident and in any other way. The cause of his illness and of his altered condition, even after the lapse of several years, was the mental shock, call it fright or what we may, which the whole circumstances of the accident wrought upon him. Wherein the changed condition of this man lies it is impossible to say, though there is no reason to believe that it is due to any pathological change, such as the unaided eye might see upon the *post-mortem* table. Certainly he is not the subject of a meningo-myelitis of the brain and spinal cord. Whether there were any peculiarities of temperament which predisposed him to nervous shock we cannot say, but it must be observed that the accident happened to this man at a time of life when the effects of shock are likely to be prolonged and severe. Happily they are still subsiding, and from a later account we learn that since he began regular work he has continued more markedly to improve.

It is impossible to refer at length to the different effects of shock upon young, upon middle aged, and old. The subject is dealt with fully by Mr Furneaux Jordan, and to his Essay we must refer our readers. Our own experience of the effect of severe mental shock in railway collisions is in harmony with his (op. cit., p. 35), where he says that, "the person with old joint disease, worn to mental and bodily torpor, and the young child, whose force is developmental rather than nervous or muscular, bear operations and injuries better than a man in the prime of life, whose every organ and function are subservient to the exercise of nerve force. In such a man the nerve force is most predominant; if such a man receives an injury the nerve force is reduced to a condition of the greatest torpor. Shock is essentially a depression or metamorphosis of nerve force. Where nerve force is predominant shock also becomes predominant." Certain it is that at the two extremes of life we have never seen such serious after-results of nervous shock from railway collision as in those who are in their prime. It must, however, be remembered that persons in the prime of life form the great majority of the travelling public, and that they furnish more examples of shock to the nervous system than do the young or old.

Evidences of the immediate effects of fright alone are, of course, not often obtainable. In the following case, however, it was recorded in the official report of the accident that "a man, name unknown, was so frightened and trembled so bad that he had to be detained all night." B. J—, æt. 44, a thick-set, somewhat robust-looking man, was in a carriage which ran off the line when the train had just left a station, and which, after jolting along off the rails for a few yards, was turned over on its side. He says he was far more shaken, "terribly shaken" was his phrase, by the previous jolting than by the overthrowing of the carriage; and when he had got out, his condition was that mentioned in the official report. On the following day he travelled home alone, presenting on arrival so dazed an appearance that his doctor was immediately sent for. When we saw him ten days afterwards he was suffering from muscular pains, increased by movement, in various parts of the body, and due, no doubt, either to bruising or straining when the carriage had been jolted and overturned. He can hardly get any sleep, having before his mind a constant *fear* of the railway accident, and he becomes occasionally "light-headed" at night. He is lying in bed with his eyes closed and the blinds down, complaining that he dreads the light. He gets very low-spirited, and frets about his business, the thought of which pains his head. He is much alarmed at the pains which he suffers, says he is afraid to move on account of them, and that he fears he has received internal injury. The bowels are confined. His temperature is 99° F., and his pulse is 102. Notwithstanding his expressions of fear, he was able to sit up in bed without sign of suffering, and in talking he moved his head naturally from side to side. He very soon also seemed quite content to have the blind drawn up, and gradually opened his eyes. There was no evidence of his having received any bodily injury other than muscular bruising and strain, and his condition was regarded by all who saw him essentially as one of general nervous shock likely to pass away after a time. We saw him again in two months. He had then a somewhat worn and anxious expression, but said he was better, his "nervousness" being not so great as it was. He complains of being easily upset and startled, and that, when thus startled and upset, there comes on a sharp pain in the head. The muscular pains are better, that which still troubles him most being a pain in the muscles of the left side of the neck. He sleeps better, though he occasionally has disturbed nights. He could walk two or three miles perhaps, but would be very fatigued. His pulse is 100. He had evidently much improved, and it was advised that after further change he should begin his work. Several weeks more elapsed, and we then found him neither looking nor feeling as well as before. He was very nervous about himself, feeling unable to do his work, depressed and melancholy, and losing heart from the thought that he would never get well. He had been attending to his business for two or three hours a day, and the anxieties of it were very distressing to him. He was, moreover, very anxious to arrange his pecuniary claim for compensation as both he and his doctors felt that that was now beginning to prey upon his mind. In bodily health he seemed well. Eighteen months after his claim was settled, we again had the opportunity of seeing him. He was then in perfect bodily health, able to follow his occupation as usual, and to endure as much physical exertion without fatigue as before the accident. He could not, however, remain so long at his desk without feeling worried, and from his wife we learned that he was more irritable than he used to be. In these respects, nevertheless, he was admittedly improving, and he himself felt confident that before long he would be absolutely well.

A case of lesser severity is the following:—M. F—, æt. 44, a man, to use his own expression, of "excitable temperament," was in a sharp collision which the whistling of the engine had warned him was going to happen. He had thus been able to prepare himself for the crash. He was conscious of having a blow on the back of the neck. He did not, however, think that he had been hurt, and was able at once to help the other passengers. This work over, he walked a mile to catch a train at another station, finished his journey, and completed the business which had called him away from home. On the following day he felt "queer" and sent for his doctor, who found him agitated and depressed, unable to occupy himself, and complaining that he felt shaken. His pulse was, however, natural and his temperature normal. He subsequently suffered from sleeplessness, and he had pains in various parts of the spinal column, where it was supposed he had been sprained or bruised. There was never any evident disturbance of the organic functions, but for some months he suffered from sleeplessness, from much depression of mind, from inability to occupy himself, and from a constant sense of weariness in the small of the back. After change of air he found himself better, and then complaining that want of work was distressing to him, he attempted to resume his business, but found that it made him worse, more nervous and depressed, and more sleepless at night. Further change of air, however, did him good, and in nine months he arranged his claim. In twelve months he was again at his ordinary work, having "entirely recovered" from his illness. He continued at his business for some years after and then retired.

These examples give as good an idea perhaps as can be given of the history and class of symptoms which cases of general nervous shock usually present. They are examples free, as we believe, from the taint of conscious exaggeration or imposture, but it must be abundantly obvious how largely the reality of many of the symptoms, lacking all vestige of objective sign, depends upon the veracity and good faith of the patients themselves. On this very account it is that there are no cases so difficult to describe, or of which it becomes so well-nigh impossible to convey an adequate impression to those who may have never seen them. It will be well, therefore, to bring together the various symptoms which are commonly met with, or of which we hear the patients complain, in genuine cases of protracted nervous shock, whether that shock has been due to bodily injury, excluding concussion of the brain, or where the bodily injury has been but trifling and the mental shock severe. We shall speak of the symptoms in the order of their frequency as gathered from a careful survey of a large number of cases, making such remarks upon each as its value or importance demands.

Sleeplessness.—It is unnecessary to dwell on the physiological value of sleep, whether we look upon it as nature's happiest means of giving rest, or as an indication that no bodily or mental disturbance is there to prevent it. Inability to sleep becomes a sign of considerable import in estimating the amount of injury and of upset which the nervous system has sustained. It shows that something has happened to break the most regular habit of life, and to interfere with the balance between healthy and disordered function of the whole nervous system. Depend upon it that the man who can sleep naturally and well after a railway collision has not met with any very serious shock to his nervous system; and that, on the other hand, returning healthy sleep, when it has been long absent or disturbed, is a very sure sign that the nervous system is regaining its equilibrium and tone. It must not be forgotten, however, that want of sleep may as surely be a consequence of other causes than the original nervous shock, and that it may form a prominent complaint in those cases where convalescence is being prevented or retarded by circumstances of which we have to speak at a future page. Thus its value is to some extent lessened as a diagnostic symptom of general nervous shock from railway collision, unless it be at periods not remote from the time of accident; and it is lessened in value, moreover, from this fact also, that we have frequently no means of knowing whether sleeplessness be undoubtedly present, for the reality of its existence may depend solely upon the statement of the patient himself.

Disturbances of the circulation.—We have said that natural sleep is an unfailing sign of the absence of serious bodily or mental disturbance, and that sleeplessness *per contra* is an indication of something having occurred to break the most constant habit of life, something whereby the healthy balance and tone of the nervous system have been, or are being, disturbed. Disorders of the circulation, whether of the heart itself,* or of more peripheral parts of the circulatory system, play a not less important part in the nervous derangement, and are very commonly to be met with in cases of general nervous shock. It has been pointed out already that the shock originally showed itself by some degree of cardiac paresis, by smallness, feeble-

* See Cases 182, 186, Appendix, and many others.

ness, or slowness of the heart beat. It is a natural consequence, therefore, that in the more serious cases of nervous shock, especially where great mental alarm tends to perpetuate the nervous disturbance, derangements of the circulation should be frequent and long-continued. The patients complain of palpitation, and palpitation from altogether trifling causes. The cardiac innervation may be so disturbed as to induce great frequency of the pulse, which may vary from 100 to 150; but far more commonly we shall find that the palpitation is occasional, and that it is only from exciting causes that the pulse beat is increased. It is important to remember this in the examination of patients, for if we count the pulse only at the beginning and omit to count it at the end of examination, we may be led to believe that the cardiac disturbance is more serious than in reality it is; and by the opposite error we may fail to discover any cardiac disturbance at all. A perfectly steady pulse throughout the whole examination tells its own tale. Nay, the rate, and character, and excitability of the pulse form an almost metrical indication of the amount of disturbance of the nervous balance, strength, and tone, and the pulse is often the only sign we have to guide us to a right estimate of the patient's condition. It is important, however, that we should learn as far as possible the character of the patient's circulation before the accident, and the existence of any constitutional states or diseases, of which gout is perhaps the most common, which may give rise to functional cardiac disorder.

But while the state of the pulse may form the test of what we may call the grosser circulatory disturbances, there are yet other symptoms which are by no means uncommon. The whole vaso-motor system may be deranged; and when we hear the patient complain of alternate sensations of heat and cold, or of flushing of the face and head;* or when we find that at one time his hands and feet are

* See Cases 24, 96, 156, Appendix.

unnaturally warm, and at another unnaturally cold, we have evidences it seems of disturbance in those peripheral parts of the circulatory system, which are not necessarily or immediately under the same nervous influence or control as the heart itself. The functional strength, if we may so put it, of the vaso-motor system has been weakened; it has lost its tone or healthy balance, and the loss is shown by the symptoms which have been named. We shall again have to refer to these disorders in another place, for they have some share, we think, in giving rise to the abnormal sensations of which we often hear, and whose very obscurity tends to cast doubt upon the *bona fides* of the man who feels them, and who can only describe them as they seem to him. With returning health and strength these various symptoms disappear, but we shall have to point out again how they are liable to be maintained by those influences and circumstances which tend to retard convalescence.

Headache.—Intimately connected with the foregoing symptoms is the complaint of headache, rarely amounting to actual pain, unless indeed there has been concussion of the brain. "Pain" is, however, so relative a term that it is difficult to estimate it at its true value, depending so largely as it does upon the idiosyncrasy of the individual. More common than actual pain in the head is the sensation of weight or oppression, coming on without evident cause, or induced by attempts at mental occupation, by agitation, or want of quietude. These sensations are doubtless dependent in a great degree on the irregularities of circulation which have been named in the last paragraph; and the exciting causes of palpitation, or of alternate sensations of heat and cold, may at the same time indirectly give rise to pain and morbid sensations in the head. Hence, also, have origin the sensations of giddiness and swimming in the head when the patient suddenly rises from the recumbent posture, sensations not uncommonly experienced by those who are first beginning to move about after serious and weakening illness. These various abnormal sensations are, moreover, largely due to the sleeplessness which is so common, and which invariably shows itself in impaired brain power so long as the brain is deprived of natural rest. Occupation of the mind very early induces brain fatigue, and this fatigue is revealed to the patient by pain or oppression of the head. And as long as the general prostration leads to lessened bodily activity, derangements of digestion, constipation, and the like, tend in a still further degree to make headache a not uncommon complaint from those who are suffering from general nervous shock. Bring about sleep and natural rest, improve the cardiac tone by restoration of the general health and strength, and it will be found that the headaches and the brain fatigue soon pass away.

Nervousness.—Again we have a phrase and symptom most indefinite in character, and one about which we have often to be content with the statements of the patients themselves. Complaints of being easily startled, of a sense of depression and melancholy, of trembling under excitement, of a desire to be alone and to avoid all noise, of hopelessness as to future prospects and the possibility of recovery, of agitation in the presence of others, of globus hystericus; of these we often hear: and beside them we may place in the same category sighing and panting, screaming at night, irritability of temper, stuttering and stammering, feebleness of voice, and the other hundred and one complaints and symptoms which we may take as evidence of nervous prostration and loss of tone, of the patient having been reduced to a more or less emotional or hysterical state, wherein loss of control is a prominent feature, whether it be as cause or effect, of the strange condition in which the patient seems.

Excessive sweating is happily a symptom somewhat less

vague than those of which we have just spoken. It is allied to the disturbances of the peripheral circulation and, like them, points in all probability to some neurosis or loss of healthy tone of the vaso-motor system. It is an evidence of impaired nerve function, and is a symptom, when it occurs without obvious exciting cause, of weakness and prostration of the patient who suffers therefrom. It is not a disease of the sudoriparous glands, but is a result of disordered and weakened function of the nerves which control their action. In a suggestive paper, "Remarks on the Mechanism of the Secretion of Sweat,"* Dr Handfield Jones, after an examination of the conditions in which sweating occurs :—in exercise where there is an "excessive expenditure of nerve force" and when the increased blood-flow will extend to the sweat glands and excite them to increased action; in the sleep of debilitated persons when there is " non-development of nerve force;" in great heat where the tendency is to "abolish nerve force;" in general debility, or in rickets where general debility is strikingly apparent; in impending syncope, and the prostration caused by tartar emetic; and sometimes in influenzal catarrh—conditions wherein "nerve force is for the time greatly depressed, while sweating is profuse,"—comes to the conclusion "that in the great majority of ordinary instances of sweating the process is essentially one of vaso-motor nerve paresis." And there can be little doubt, we think, that in the cases of general nervous shock after railway collisions excessive sweating, whether general or local, points to weakening or paresis of the vaso-motor system, whereby the sudoriparous glands are liable to morbid activity of function from lack of due nerve-control.

Here, also, in all probability, we should rightly include those vaso-motor disturbances which, limited to special regions or tracts, may also give rise to very obvious symptoms. Of these polyuria is the most pronounced; but

* 'Journal of Anatomy and Physiology,' vol. xv, p. 238.

we have seen several cases where profuse diarrhœa has occurred almost immediately after a railway collision, and menorrhagia is a by no means infrequent consequence. All these must be regarded as kindred signs of vaso-motor disturbance, and they open up this important question, how far they may depend upon some primary disturbance or paresis of the vaso-motor centre—if there be but one—in the upper part of the spinal cord, or of the vaso-motor centres—if there be many—in different parts of the cerebro-spinal system. Why this vaso-motor centre, or why the nucleus of the pneumogastric, should be especially liable to derangement, it is impossible to say; but it must be borne in mind how closely the varied functions controlled by these centres are allied to the emotions, and the morbid exhibition thereof. One of the most remarkable cases which we have ever met with is that of a woman who was greatly alarmed at a railway station by an accident which she thought was going to happen to her child, and who was herself knocked down on the platform. She had some years before suffered from polyuria, and this again came on shortly after the accident. The polyuria continued for some weeks, with one exception of twenty-four hours during which it was replaced by an enormous flow of milk from the breasts, the vaso-motor disturbance seeming to move from one part to another. She was not at the time suckling. Cases such as this seem to support experimental observation as to the presence of vaso-motor centres in the cerebral hemispheres, and better to explain the intimate association of emotional and vaso-motor disturbances than if the centres lay in the spinal cord alone.

Asthenopia and size of the pupil.—Leaving altogether out of account those cases, happily in our experience exceedingly few in number, where there has been actual damage to the ball of the eye, we find that a by no means uncommon complaint is of some defect of vision. "I can read for a short time, and then the lines all seem to run together," the patient tells you; and he thus describes a symptom, like those which have gone before, of prostrate nerve force. The asthenopia is due in nearly all cases—whether there has previously been ametropia or not—to loss of accommodative power, as a result of the general weakness and depression which render any sustained effort difficult or impossible. It is merely another sign of easily induced fatigue; and in those persons who have neither error of refraction nor presbyopia, the asthenopia will disappear—as the general muscular fatigue and the brain fatigue will disappear—with returning health and strength. In all probability it will not entirely pass away in those who have some anomaly of refraction, brought to light for the first time by the prostration of the nervous shock. The asthenopia is precisely the same as that which is a not uncommon symptom of hypermetropia, and which may be felt for the first time after an exhausting illness or during the weakness induced by prolonged lactation. In lusty health the accommodative power was adequate to overcome the refractive error and to prevent any fatigue of vision; reduce the strength, and asthenopia is revealed as a direct consequence of weakened power of accommodation. It is very doubtful whether this power will ever be perfectly restored to its original strength in those who have abnormal refraction, or in whom presbyopia is either imminent or advanced. Thus we find it occasionally happen that persons who have suffered from the shock of a railway collision need in future years to wear glasses in order to correct the refractive error of which they had not been conscious before. It is in these cases of induced asthenopia that muscæ volitantes are so commonly seen, and so often alarm the patient. They are of no pathological significance whatever. "They are dependent on microscopically small bodies, which in every one float in the vitreous humour;"[*]

---

[*] 'Accommodation and Refraction of the Eye,' Donders, p. 394.

and we may add that there are singularly few persons to whom they are not at times apparent.

The eye presents one sign, which occasionally may be the only objective indication of loss of nervous tone—we refer to the size of the pupil. "Speaking generally, it may, I think, be asserted," says Mr Hutchinson, "that there is a relationship between the size of the pupils and the state of the patient's nerve tone, due allowance being made for age, and the other circumstances just mentioned (sex, age, refractive errors). If the tone be low the pupils are large. .... The dilatation of the pupil which goes with a feeble state of circulation is due to defective tone of the retina and nerve centres, rendering necessary a large supply of light. It may also be due in part to impaired tone in the circular fibres of the iris. No condition of defective tone is more certainly revealed by large pupils than that which results from sexual irregularities in early life."*

Our observation of the state of the pupil in cases of general nervous shock after railway collision fully coincides with the teaching of Mr. Hutchinson, whatever the physiological explanation may be.

Loss of memory.—It is strange how common is the complaint of "loss of memory," and yet the phrase is an exceedingly inappropriate one to convey an accurate description of that which the patients usually mean thereby. The "loss of memory," of which we have here to speak, is not an inability to recall the events and incidents of past life, but rather an incapacity for sustained thought, and for continued application to the work which may be taken in hand. It is a lack of the power of volitional attention, and is a symptom of easily induced fatigue. It is not a symptom of serious import, nor is it an evidence, as we have heard suggested, of some mysterious mischief

* 'Brain,' vol. i, p. 5, "Notes on the Symptom-significance of different States of the Pupil."

in the brain. It is another phase merely of the general weakness, and of the inability to apply himself to any settled occupation, which a patient very naturally feels under the sense of weakness and depression incidental to the nervous state in which he is. True loss of memory is very rare, and there is hardly any complaint which we shall find so full of inconsistencies as this one, or where it is more desirable to recognise the true import of the term.

Amongst the earlier manifestations of functional disturbance from shock must be mentioned suppression or other derangements of the catamenia.* And of not less importance because it may be serious, if undiscovered, in its after-consequences, is retention of urine as a direct result of shock. We have seen more than one case in which retention of urine with hyper-distension of bladder has led to most troublesome atony of bladder and incontinence, a symptom which has thereupon been erroneously attributed to "concussion of the spine," and some mischief in the spinal cord.

It is a remarkable circumstance that railway collision should so rarely produce abortion or premature labour. The same observation has been made by others. Although many cases have come under our notice of pregnant women suffering from "nervous shock," in not one has there been miscarriage or any apparent derangement in natural labour. Surely this is strong evidence that the *concussion* is not so great as it is often imagined to be, for violent disturbance of the uterine contents is one of the well-recognised causes of induced labour.†

We have thus gathered together a number of symptoms, and a larger number of the complaints, which we meet with in cases of prolonged nervous shock. They comprise the prominent features of the illness which may supervene, or continue, after the collapse and its more immediate results

* See Cases 130, 165, 177, Appendix.
† See Cases 118, 138, 157, 230, in Appendix, and many others.

have completely passed away. To the immediate results themselves we shall make no further reference, for the stage of collapse and the stage of reaction differ only in degree, and chiefly in the degree of emotional disturbance, from the collapse and reaction which are met with after other injuries received in other ways.

In a subsequent chapter we shall deal with some of the graver consequences of profound emotional shock, with the convulsive seizures, and with the various neuromimeses, whose development forms so interesting and important a feature in the history of nervous shock from collision injury.

We have spoken of the symptoms detailed as the symptoms of protracted nervous shock, whether due to injury or to purely mental causes, and we have excluded from our inquiry the symptoms which arise from concussion of the brain. These have been so admirably tabulated by Mr Hutchinson,* than whom no man has had larger opportunities of seeing and observing concussion injuries of the head, that we shall do well to quote his table in full, and more especially because the physiological explanation which he therein offers of the symptoms will suffice to show that the symptoms of general nervous shock are very closely allied to those which we meet with in cases where there has been unquestionable collapse from concussion of the brain. (See end of the chapter, pp. 186, 187.)

We would especially direct attention to the physiological explanation which Mr Hutchinson gives of the various phenomena seen in the earlier stages of concussion collapse, for there can be little doubt that the same causes produce a great number of the symptoms which are met with in the cases with which we have here more especially to do, and which rather fall into the fourth stage of Mr Hutchinson's table.

* 'Illustrations of Clinical Surgery,' vol. i, p. 86.

The fourth stage, then, he defines as one of gradual convalescence, lasting for an "indefinite period," in which some of the "symptoms may often remain for years as the sequelæ of *severe concussions*." (Italics our own.) "The symptoms are essentially due to imperfect recovery of the vaso-motor system and easy production of local turgescence of vessels." It is essentially in this, the fourth, stage, or stage of gradual convalescence, that the symptoms of general and protracted nervous shock as the result of combined bodily and psychical causes are to be found after railway collisions. While, however, the great majority of cases may happily be placed in the stage of convalescence, there are occasional instances of persons who never reach it, having succumbed at an early date to the severity of the shock. No case of this kind has fallen under our own observation, but the history of nervous shock cannot be complete if we exclude the following examples of death from general shock, which have been communicated to us by a distinguished surgeon of vast experience in railway injuries, the only cases we may say which he has met with in a period of more than thirty years.

The first case is that of a man forty years of age, of exceedingly delicate physique, who was in a collision at night. The accident was a slight one, and he was the only passenger injured. He was said, in the official report, to be "violently shaken," but he was able to go on home. The next day he was delirious, and on the third day he was still talking somewhat incoherently. He complained of being much shaken and of feeling seriously injured, but there was no evidence discoverable of bodily hurt. He improved for a time, and his condition was not thought to be serious. He never seemed, however, to make any marked progress, and four weeks after the accident he became more prostrate, and greater anxiety was felt about him. From this time he gradually got weaker and weaker, and died on the thirty-seventh day. No organic disease whatever was found on post-mortem examination in any of the viscera. The lungs were greatly congested, and the cavities of the heart were distended with blood, as if death had occurred from failure of respiration and circulation.

The accident was regarded as the unquestionable cause of the death. Though moderate in character, it no doubt exercised a very unusual influence in depressing vital powers—never very strong naturally—and finally inducing such an amount of nervous exhaustion as to terminate fatally, even though there were no evidences before, or after, death of physical injury to any one part.

The other case was that of an apparently strong and healthy girl, nineteen years of age, in good position in life, who was in a most serious collision. She received no bodily injury, but on the night of the accident she woke screaming that the engine was rushing into the room. Her illness followed much the same course, and she died in about five weeks, no structural disease whatever being found after death. The brain and spinal cord were examined in both instances.

Cases such as these are not peculiar to railway accidents, and similar examples of death from mental shock without any organic change discoverable are recorded in works on nervous diseases.* They have prompted a question, which has not yet been answered—whether death may not be the result of some effect produced upon the blood itself, whereby the natural processes of nutrition are arrested, and life comes to an end.

"It is not improbable," writes Mr Jordan,† "that further knowledge will discover some specific cause of death in many of these cases; but it is also equally probable that many cases will remain which can only be regarded as cases in which shock is not instantaneously fatal, but only gradually though uninterruptedly passes to a fatal termination."

* See Wilks, op. cit., p. 397. † Op. cit., p. 41.

TABULAR STATEMENT OF THE STAGES OF CONCUSSION, THE SYMPTOMS PRESENT IN EACH, AND THEIR PHYSIOLOGICAL EXPLANATION.

DEFINITION.—*Shake of the cranial contents without any structural lesions of importance.*

Stages.	Duration.	Phenomena and their physiological explanation.	Physiological explanation.	Remarks.
1st stage: COLLAPSE. Chief feature paresis (cerebral) of the heart.	A few minutes, an hour or two, or even one or two days. In the majority not more than an hour, may end in death.	Insensibility. Reflex irritability suspended. Pallor. Cold surface. Pupils neither dilated nor contracted. Feeble pulse. Feeble respiration.	Nervous centres not supplied with blood, and suffering directly from "shock." Cutaneous capillaries empty. *a.* No blood in the skin. *b.* Nervous supply suspended. Temporary shock; paresis of third and vaso-motor nerves. *a.* Temporary cerebral paresis of the heart. *b.* Spasm of arteries (?). Medulla oblongata and spinal cord anæmic, and suffering from "shock."	This stage always begins from the moment of the fall. If there has been a period of consciousness first then hæmorrhage is to be suspected.
2nd stage: RALLYING (vomiting stage). Recovery of the heart from its temporary paresis, with partial relaxation of arterial walls.	A few minutes or a few hours, rarely longer than twelve hours.	Partial sensibility. Reflex irritability partially restored. Return of warmth and colour. Vomiting. Better pulse and respiration. Mental distress and uneasiness. Subjective cold, with shuddering.	More blood supplied to brain. More blood to brain, medulla, and spinal cord. Cutaneous capillaries better filled. *a.* Previous arrest of digestion. *b.* Recovery of pneumogastric nerves from paresis. Recovery of heart from temporary paresis. Irregular and, as yet, insufficient supply of blood to different parts of the brain. *a.* Supply of blood still imperfect.	Vomiting is very often the symptom which ushers in this stage.

SHOCK TO THE NERVOUS SYSTEM 187

...ry] stage" (full recovery of the heart's power; paresis of the arteries).	There is no risk of death in this stage unless, in addition to the general concussion, local contusions have been received, about which softening may occur.	Great sleepiness, but may be aroused.	Doubtful.	...sent they are to be referred to local lesions and not to the concussion proper.
		Consciousness when awake. Occasional restlessness.	Brain not structurally damaged. Certain parts of the brain irregularly supplied with blood (possibly contused). Paralysis of arteries.	
		Large, slow, full pulse. Intermittent pulse.	Rhythm of heart disturbed by shake of the medulla.	
		Surface hot and flushed.	Full supply of blood to skin on account of paralysed arteries.	
		Surface dry.	Activity of sudoriparous and sebaceous glands in abeyance.	
		Tongue dry or dryish.	*a.* Patient sleeps with mouth wide open. *b.* Activity of buccal and salivary glands in abeyance.	
		Bowels constipated.	*a.* Paresis of intestinal walls. *b.* Non-secretion by intestinal glands.	
		Bladder tolerant of distension.	*a.* Paresis of its muscular walls. *b.* Diminished sensibility of nervous centres.	
		Pupils contracted when asleep, and normal at other times. Mental faculties confused. Irritability.	Recovered function of vaso-motor and third nerves. Passive congestion of brain. Irregular circulation of blood in brain.	
4th stage: GRADUAL CONVALESCENCE.	Indefinite.	Gradual return of memory and other mental faculties. Irritability of temper and tendency to outbreaks of passion. Susceptibility to the influence of alcoholic stimulants. Headaches.	Imperfect recovery of the vaso-motor system and easy production of local turgescence of vessels.	Some of the symptoms mentioned in this stage often remain for years as the sequelæ of severe concussions.

CHAPTER V

SHOCK TO THE NERVOUS SYSTEM (*continued*)

IN the previous chapter we have endeavoured to bring together the symptoms and signs of nervous shock ; but when their aggregate has been told, and their physiological import explained, a faint and but imperfect outline is drawn of the deplorable state which patients of both sexes, men not less frequently than women, may present when suffering from this condition. "Men not less frequently than women ;" for even if in every-day life women more commonly than men show signs of being emotional, excitable, and hysterical, it is nevertheless true that, as a direct outcome of the nervous shock of a railway collision, men become no less emotional and hysterical than they. We are much inclined to agree with Mr Jordan[*] that "the frequency of hysteria (if such a term may be used) in men is not fully recognised ;" but if the manifestations thereof, as we may admit, are absent or but rarely seen in ordinary men, a condition closely allied to the "hysteria" of women is very common, or is commonly developed, in men, after the great psychical shock of a railway accident. We know no clinical picture more distressing than that of a strong and healthy man reduced by apparently inadequate causes to a state in which all control of the emotions is well-nigh gone ; who cannot sleep because he has before his mind an ever-present sense

[*] Op. cit., p. 27.

of the accident; who starts at the least noise; who lies in bed almost afraid to move; whose heart palpitates whenever he is spoken to; and who cannot hear or say a word about his present condition and his future prospects without bursting into tears.

We have said that the class of symptoms which we have been considering ought to be placed—and we believe with every right and justice they should be placed—in the fourth stage of Mr Hutchinson's table, or in that of convalescence; for however sad may be the circumstances of individual cases, it is a fact that in the large majority the tendency to recover is very strong. The third stage at an end, the fourth or stage of convalescence is entered upon; and the symptoms, serious perhaps at first, gradually assume a type of lesser severity, and in the course of a few weeks or months they pass almost imperceptibly into a state of health, and the patient is able once more to resume his business and to engage in the ordinary pursuits of life. Exceptions there doubtless are, as in the first case which was related (see p. 167); but fortunately it is rare to meet with examples of such marked and long-lasting damage to the stability of the nervous system.

But although the tendency towards convalescence and complete recovery sets in in a great number of cases at a very early date, there are others in which convalescence is unduly delayed; unduly, because the symptoms and their duration seem out of all proportion to the original injury or shock. There are yet other instances where convalescence may have advanced almost to recovery, where the patient has manifestly improved day by day and has been almost well enough to resume his work, and yet in whom the symptoms recur in all their severity, and the period of convalescence may be very much prolonged. It behoves us very carefully to inquire into the reasons which conduce to this protraction of the illness, and which conduce also in great measure to delay in convalescence, when all the circumstances, the amount of injury, and the evidences of initial shock, pointed in the direction of very early recovery.

We have spoken very little hitherto of the bodily injuries received, and have assumed throughout that the causes originally at work to bring about the shock and its after-consequences were essentially psychical. The fact, however, must never be lost sight of, that there are few cases of nervous shock after railway collision in which *some* bodily injury has not likewise been sustained. The mode of accident, as scores of cases abundantly testify, has an unquestionable tendency to cause injury of the vertebral column, an injury which in the great majority of cases is a simple sprain of the spinal muscles and spinal ligaments, with the inevitable consequence of severe vertebral pain. Sprains are, moreover, not unlikely to have been received in other parts of the body, even when the patient was perfectly conscious at the time that he had no blow, and not a mark is subsequently to be seen. Hence it comes to pass that from the inherent nature of the bodily injuries themselves, pain in various parts of the body—in the trunk and in the arms and legs—is very liable to come on some time after the accident, to be severe in character, deep-seated in position and, from the absence of all bruising, seemingly most obscure. Psychical elements again rise to aggravate the patient's condition. His mental balance has already been upset by the shock of the collision, and it is disturbed still further by the onset, the character, and the obscurity of the pains which supervene. And it is obvious that this result is most likely to happen in those cases where the appearance of the pains has been delayed, as is not uncommon, for two or three days. They renew the alarm of the sufferer, whose attention is thereby more closely directed to them, and their import becomes gravely aggravated in his mind.

These pains, moreover, are prone to increase in severity during the first few days and to last for a long time, and their very duration tends to maintain the exaggerated estimate which has been formed of them by the patient himself. Nor does familiarity with them lessen his alarm, for the original psychical disturbance has laid the sure foundation for an altogether erroneous estimate of the sensations which he feels. And thus we find that before very long the mind of the patient, unhinged by the shock, and directed to the pains and other abnormal sensations of his body, tends as it were to run riot with the symptoms which he feels. Dwelling constantly on his bodily sensations, he is on the look-out for any new sensation that may arise, and is alive to and makes discovery of sensations which to the healthy have no existence at all. Is it possible, we would ask, that a large number of the abnormal sensations of which patients so frequently complain while the mental balance and tone are thus perturbed, can be due in any measure to a conscious perception of the perennial sensations of organic life? The "hysterical" condition—we use the word for want of a better and without a shadow of reproach—the hysterical condition is essentially one in which there is loss of control and enfeeblement of the power of the will, and amidst the various ways in which these may show themselves, there is loss of the habitual power to suppress and keep in due subjection the sensations which are doubtless associated with the various functions of the organic life of the individual. In the process of evolution towards a higher state of intellectual activity and endowment man, we take it, has become more and more unconscious of the sensations which of necessity accompany the functional activity of the various organs and structures of his body. That the stomach, for example, the liver, the heart, the ovary, the œsophagus, are, as are the organs of special sense, represented somewhere and somehow, though in less degree, in the sensorium, is highly probable on *a priori* grounds, and is, moreover, established by the experiments of morbid physiological activity, by the abnormal sensations, *e.g.* which, as the *aura*, may affect these and other parts at the commencement of an epileptic discharge. Dr Maudsley even takes " it for granted that each internal organ of the body has, independently of its indirect action upon the nervous system through changes in the composition of the blood, a specific action upon the brain through its intercommunicating nerve-fibres, the conscious result whereof is a certain modification of the mood or tone of mind. We are not directly conscious of this physiological action as a definite sensation, but none the less its effects are attested by states of feeling that we are often perplexed to account for."*

And if in perfect health of body and stability of mind these varied sensations play little part in the sentient life of the individual, it is because the intellectual development of man has enabled him to control them and to allow them neither lot nor share in the sentient consciousness of active life. In the lower animal, whose brain is hardly differentiated from the other parts of its nervous system, or which has no brain at all, the organic sensations doubtless have a more important part in the economy, and probably in the enjoyment of life ; but as we step higher and higher in the scale of development, with increasing size and complexity of brain, the organic sensations have a proportionately smaller representation in the centres of intellectual activity. Let some sudden profound psychical disturbance arise, such as may well be induced by the shock of a railway collision, the intellectual control at once is lessened and the organic sensations declare their being, and force themselves into the conscious life of the individual. " If the nervous system," writes Sully, " has been slowly built up, during the course of human history, into its pre-

* 'Pathology of Mind,' p. 30.

sent complex form, it follows that those nervous structures and connections which have to do with the higher intellectual processes, or which represent the larger and more general relations of our experience, have been most recently evolved. Consequently, they would be the least deeply organised, and so the least stable; that is to say, the most liable to be thrown *hors de combat*.

"This is what happens temporarily in the case of the sane, when the mind is held fast by an illusion. And, in states of insanity, we see the process of nervous dissolution beginning with these same nervous structures, and so taking the reverse order of the process of evolution. And thus, we may say that throughout the mental life of the most sane of us, these higher and more delicately balanced structures are constantly in danger of being reduced to that state of inefficiency, which in its full manifestation is mental disease."*

And thus it is, it seems to us, that when by the profound shock of a railway collision the " higher intellectual processes " are thrown *hors de combat*, these organic sensations which, as the same writer says, " constitute for the most part in waking life an undiscriminated mass of obscure feeling, of which we are only conscious as the mental tone of the hour," and which form " ' as the vital sense ' an obscure background for our clear discriminative consciousness, and only come forward into this region when very exceptional in character,"† step out of their natural obscurity, and become the foci of the uncontrolled and misdirected attention of the mind.

"In diseased states of the nervous system variations of sensibility become much more striking. The patient who has hyperæsthesia fears to touch a perfectly smooth surface, or he takes a knock at the door to be a clap of thunder. The hypochondriac may, through an increase

* 'Illusions,' p. 122.
† Op. cit., pp. 148, 145.

of organic sensibility, translate organic sensations as the effect of some living creature gnawing at his vitals."*

"When the hypochondriac complains unceasingly of the distressing anomalous sensations in his interior, it is a question whether he has not cultivated such a hyperæsthesia of his organic sensibilities by constant attention to them as to be rendered sensitive to the functions of his organs or even to the passage of food through the intestines."†

Be this, however, as it may, and be the explanation what it may, as a matter of fact we see the results of fixed attention, expectation, and idea in their most exaggerated forms in these cases of general nervous shook.‡

Hence arise in great measure the sensations of creeping and crawling, of tingling and burning, of oppression and weights, of numbness and deadness, and of approaching palsy; the hyperæsthesia and the anæsthesia; the gnawing and never-ceasing pains; the startings, tremblings, and spasms; the pantings and sighings; the fear of moving; the melancholy of the present and the hopelessness of the future, which are such prominent symptoms of the emotional state.

* Sully, op. cit., p. 65.
† Maudsley, op. cit., p. 369.
‡ See a remarkable case recorded by Mr Wrench, of Baslow, in the 'Lancet' of January, 1880, p. 71. "A gentleman, away from home, suddenly while dressing found his mouth and nose full of blood, and, at the same time, became aware that his false teeth, which he seldom removed at night, were missing. He fancied that he felt them in the pharynx, and sent anxiously for surgical aid. Examination failed to detect any foreign body, still the patient persisted in saying he had swallowed them, and complained of intense pain as if the region near the hyoid bone was touched with a probang, asserting that the teeth were there. Finding that the patient could swallow fluids, Mr Wrench instituted a search for the teeth, and, fortunately, they were found on the top of a chest of drawers. All symptoms immediately vanished, and the patient dressed and ate an excellent lunch."

That these psychical disturbances have some physical substratum seems highly probable. We have seen that continued disorders of the circulation are the most obvious after-signs of shock, due to the paresis of the heart and circulation. " Imperfect recovery of the vaso-motor system, and easy production of local turgescence of vessels," Mr Hutchinson tells us, is the physiological explanation of the symptoms which characterise the period of convalescence in his table (vide *ante*, p. 187) ; and there are therefore, in all probability, local disturbances of the vaso-motor mechanism in parts far distant from the central organ of the circulation itself.

Is it not, therefore, on the other hand, likely that many of the abnormal sensations of which we hear, and which form the burthen of the patient's complaints, have a real, not an altogether imaginary, basis in transient flushing, or in transient anæmia, of the affected part ? Some such cause as this must lie, it seems to us, at the root of many of the sensations which afflict peripheral regions of the cutaneous surface, and possibly of those also which are felt in the central or in more vital parts. In his remarks on the mental state of patients who afford examples of nervous mimicry, Paget writes, that "the distribution of blood is, in many of the mimic cases, greatly affected. Heat and cold of the same part rapidly succeeding one another, flushing and pallor, turgidity and collapse, all these are frequent, striking, and capricious in the nervous mimicries ; but, after months and years of their occurrence, not one organic change may be discerned. You may find in our hospital reports the case of a gentleman who consulted me because, for several years, whenever he walked far or fast, his feet became cold, white, and numb—' dead,' as they are called ; and then, when he rested, they flushed red and hot, and were turgid with blood, distending even the veins of the leg. Yet, after years of such disturbance, all the structures of his feet were as healthy as any of yours are. And I urge you," he says, "to study carefully such cases. They are rare, but you are likely to meet with them, and when you watch them think of what may happen if such changes as you see in the skin, now looking bloodless and now over-filled with blood, can, on as little provocation, happen in the brain or spinal marrow, in the heart or in an ovary. What startling and yet what harmless disturbances they might produce! what mimicries of grave disease!"*

And as the mental state may be affected and deluded by the abnormal sensations, so in an even greater degree may the abnormal sensations be affected by the mind. The results of attention concentrated on a part are seen in their most aggravated forms. "Attention," as Paget truly says, in writing of nervous mimicry, "makes all pain sharper; without it there would be little or none; and attention long exercised becomes keener, more direct and definite; and so the perception of pain becomes more intense."† And as of pain so of other abnormal sensations it may be said that prolonged attention makes them more acute and more oppressive, and that they become more dominant in the mind, and less under the control of the already weakened will. "Reflection, and especially the anxious reflection, upon any of the bodily sensations, increases them to a morbid extent, and may originate a host of imaginary disorders."‡

Small wonder that the patient, alive to every new sensation which may arise, should tend to exaggerate its import, to describe it in terms which to the healthy man seem well-nigh absurd, and that exaggeration should be a pronounced feature of the morbid state which we recognise as "hysteria," no matter which be the sex affected for

* 'Clinical Lectures,' 2nd edit., p. 184.
† Op. cit., p. 186.
‡ Tuke, 'Influence of the Mind upon the Body,' p. 135.

the time.* And out of this very exaggeration itself arises another cause of prolongation of the illness. The exaggerated estimate of the symptoms themselves leads to an erroneous estimate of the present incapacity, and to an increasing belief in the impossibility of future recovery and usefulness. It is only natural that differences of opinion should hence arise between those who have to receive compensation for the injuries and for the prospective consequences thereof and those who have to provide the compensation, and who take a wholly unsentimental and, we are afraid, sometimes an inadequate view of the value of the patient's health and life. The case drifts on, months perhaps are wasted in an altercation which ends either in the patient having materially to modify the views he held as to the compensation he ought to receive; or, worse than this, he is drawn unwittingly into litigation, and now an unhappy plaintiff he is subjected to the delays and anxieties and worries which a lawsuit involves. What surer means than this for aggravating the symptoms of which the patient complains? There is no greater service which his medical attendant can do him than by using his influence to restrain him from a course which must be fraught with unspeakable mental anxiety, and which must retard, if not altogether for the time prevent, any possible restoration of the nervous balance and tone. Speaking of the troubles which beset her father's life and her own early years, Fanny Kemble writes in her 'Record of a Girlhood' that "that dreary Chancery suit seemed to envelope us in an atmosphere of palpitating suspense or stagnant uncertainty, and to enter as an insensible element into every hope, fear, expectation,

* "The misfortune is that the disorder which strengthens the tendency (*i. e.* 'to exaggerate much and even to simulate symptoms, apart from any question of intentional deceit') weakens the will, and so leaves less power to control what is more difficult of control."— Maudsley, op. cit., p. 325.

resolution even, or action of our lives."* And no less surely does litigation keep the sufferers from nervous shock in an atmosphere of palpitating suspense or stagnant uncertainty, entering into their every hope, expectation, and fear. Is recovery possible under such an influence; is there not, indeed, every likelihood that we shall find their symptoms getting worse and worse, or at best undergoing no change, symptoms which we have endeavoured to describe, and which are now more fittingly termed " litigation symptoms," than those of general nervous shock?

And herein also lies the explanation of the great majority of those cases where improvement has advanced to such a stage that return to work seems on all hands desirable, and yet nevertheless when work is suggested or attempted improvement stops; and even in some instances the patient seems from that very moment to fall back and to become worse than he was before. For, as a matter of fact, it is very rare to find the patients return to work so long as the question of compensation, and the possible disputes attending it, remain unsettled. Now and then we may meet with a patient, over whose plans and resolves the time and matter of compensation have little influence, who returns to his business with every benefit to himself at the moment when he has sufficiently recovered to do so. Such instances, however, are the exceptions to the rule, and occur probably in those only whose mental balance has never been very seriously upset, in whom the symptoms of general nervous shock have not been severe, or who have the good judgment and adequate determination not to allow these matters to weigh upon their minds. The experience of hosts of cases establishes this fact, that patients will not, or cannot, make the necessary effort to resume work so long as the settlement of the pecuniary claim has been uneffected. And thus, in addition to the worries and anxieties of litigation and dispute, there

* 'Record of a Girlhood,' vol. i, p. 143.

arises another very potent cause for continuance of the symptoms, and for inducing a state of "chronic invalidism," which is far more dependent upon the circumstances of the moment than upon the original nervous shock received.

This powerful cause is the want of occupation. Can anything be worse for a man, is there anything more likely to lead to irritability and fretfulness, to sleeplessness and loss of appetite, to nervousness and anxiety about himself, to hopelessness as to the future, to a lack of power to concentrate his attention upon anything which he may have in hand, is there anything more fitted to disturb the *mens sana* and the *corpus sanum* than want of healthy occupation? And when this goes on, as we too frequently see it, for months and months, each month more wearisome and more wasted than the one before, is it to be wondered at that the picture which these patients present is often lamentable indeed? Still more wretched is it likely to become if, in addition to the want of occupation, the patient has remained altogether indoors, and has been deprived of the good which healthy bodily exercise might have done him.

Make all the allowance that we honestly may for the special circumstances of alarm attendant upon a railway collision—and we would not for a moment seek to lessen their real influence—and compare the state of one whom for the nonce we will call a railway patient waiting for compensation with the state, as nearly similar as may be, of a hospital patient who has had no compensation to look forward to, and who has been *compelled* to resume his work as soon as he was able, and we shall see how different is their lot, and how infinitely happier and less wretched is the one man than the other. The hospital patient has long ago been well, while the railway patient has been waiting, for months it may be, until compensation has been paid him, verily believing that he could not return to work and to a natural and more healthful mode of life.

We have remarked before that in cases of serious injury to limb, such as fracture, whether simple or compound, even if there be extreme collapse at the time of the accident, it is most unusual to meet with the protracted after-symptoms which have been described as due to general nervous shock. This is a fact about which there can be no doubt,* and it is one of very considerable interest also in that it throws light upon the nature of the symptoms which the more ordinary cases present. This seeming anomaly depends almost entirely, we believe, upon psychical causes. The very definiteness of the injury presents a point of focus for the patient's mind, one, moreover, especially suitable because the injury tends—and as far as he knows usually tends—towards recovery and restored usefulness. The collapse subsides, and the patient finds himself with an injury not more obscure, it may be, than that of a broken leg. He knows that he has been doomed for a time to his bed, and that when confinement thereto is no longer necessary he will begin to move about again and to get well. The injury is definite and precise, its symptoms are obvious from the moment it was received, it lacks the seeming obscurity that is a feature of those symptoms which only supervene after several hours or days, there is probably less pain as time goes on, and all the circumstances combine to induce a repose of mind which is absent from the commoner cases which have been considered. And there is also the necessity of complete bodily repose from the moment that the patient can be placed in bed. The enforced rest is good both for body and mind. Confinement to bed at an end, the patient is only too thankful that he is able to move about again, and gradually begin to walk. Returning strength goes hand in hand with

* See Cases 31, 32, 38, 150, 199, 203, Appendix.

returning possibility of exercise, and there is less excuse for staying in doors because of the fear that the after-consequences of some wholly obscure injury may turn out very serious. There is, moreover, less likelihood of dispute arising as to compensation, and the money calculation becomes all the easier and the readier because the nature and extent of the injury can be definitely appraised. Thus the absence of the symptoms of continued nervous shock in cases where the amount of true collapse may have been originally severe, tends to support the view that those symptoms are due to mental causes rather than to the bodily injury or to any vibratory jar sustained.

And how largely the continuance of the symptoms is due to these mental influences is shown, perhaps even more strikingly, by the often speedy recovery which ensues when the exciting causes of the symptoms are removed. It is all very well to say—and it is an easy enough diagnosis to make—that so-and-so, who recovered as soon as his claim was settled, was "shamming," and that his symptoms were altogether untrue or wilfully exaggerated; but this will hardly suffice, nor can we accept it, to explain the symptoms which have caused so much anxiety and trouble and have been so little amenable to treatment.*
The man recovers quickly because the goal, whose prospect unsettled him, has at length been reached, and because it no longer stands in the way of his making the requisite and successful effort to resume his work. "How long I have been ill, how little I improve, how small seem the chances of my recovery," have been the uppermost sentiments in

* "The gradual influence of favorable surroundings—to wit, a suitable moral atmosphere, distracting occupations, diverting amusements, a steady reasonableness of life—will exert an unconscious beneficial influence upon the uninfected mental organisation, until the large part of it which lies outside the morbid area gains strength enough to have a controlling hold of the morbid action and to bring it by degrees into subordination to the laws of healthy function."—Maudsley, op. cit., p. 203.

his mind, and they speedily give way to this one which is wholly different and far more hopeful, "How soon shall I be well." And, again, the fact of recovery shows that the symptoms could not have been dependent on structural or organic change.

Yet another cause is there of prolongation of the illness and delay in convalescence. Bromide of potassium has much to answer for in the protraction of the symptoms of general nervous shock. A remedy of most unquestionable value in the treatment of epilepsy, it is extraordinary with what little discrimination this drug is used in the treatment of almost any and every kind of nervous disorder. Not once, but scores of times, has it come to our knowledge that patients in the pitiable condition which we have attempted to describe, have been taking bromide of potassium in considerable doses morning, noon, and night for many weeks, administered under the fond belief that the poor creatures needed some *sedative* to wipe away their tears. But when the longed-for goal has at length been reached they leave their bottles of bromide behind them, and the remedial value of this powerful *depressant* is forthwith revealed. It is not by a lavish use of the bromides that success in the treatment of neurasthenia, to which many of the cases of railway shock are so nearly allied, is being obtained, even in the most extreme cases, by Weir Mitchell, Playfair, and many others.

We have often heard it stated that the bromide is given to "quiet the nerves," but if our view of the nature of these cases be correct, the "nerves" are unfortunately too quiet already. For the symptoms of bromism are essentially those of general depression both mental and bodily, irritability of temper, a continued feeling of exhaustion and incapacity for any work, intellectual dulness and loss of memory, a sense of utter feebleness, a tendency to be despondent and to cry, and loss or impairment of sexual desire. And these symptoms are very closely akin to those of general nervous shock after railway collision. It has been shown by experiment that bromide of potassium is a very powerful depressant, and moreover it is only those who are in vigorous bodily health, as Dr Anstie pointed out years ago,* who can bear its prolonged administration. Epilepsy seems in some cases to confer a sort of immunity against the risks of bromism, but it is far otherwise in these cases of nervous shock, and we cannot too strongly condemn the almost universal practice of treating them with bromide of potassium in large doses many times a day. It is no easy matter very often to procure sleep, but other soporifics are at hand, and in America bromide of sodium is largely taking the place of bromide of potassium for this purpose, and is infinitely preferable in that the sodium salts are not nearly so depressant.

It remains for us to consider to what extent recovery is possible in these cases of nervous shock, and how far the patient regains the mental and bodily vigour which he had before the accident. Happily the record of cases which we have been able to collect is conclusive upon this point that recovery is usually very complete, and the patient is able to resume his occupation and to carry on his business as well as he did before. There are, of course, exceptions to the rule—what rule has none?—which show that some alteration has taken place in the bodily physique and very possibly in the mental vigour also. Thus we may hear that the man is less able to bear prolonged fatigue, either bodily or mental, that he is more susceptible to the influence of alcohol, more irritable and easily excited, that he lacks that complete self-control which he may formerly have had in his business relations with his fellow men, that he is nervous when travelling, that he is afraid to ride or drive, and has been compelled to give up his hunting and shooting, that he is a more nervous man than he was before, that he is more subject to headaches, and in the severer cases that

* 'Neuralgia and its Counterfeits,' p. 191.

his hair has turned grey and he looks prematurely aged. Some years have been added to his life, and he is never quite what he was before.

The completeness of recovery depends largely, of course, upon the severity of the original mental shock, and to a great extent also upon the temperament of the individual. If he is of a " nervous temperament "—whatever that may mean —he is a bad subject to suffer from the shock of a railway collision, and if there be added a gouty constitution, then his chances of extreme suffering therefrom are greater still. The evil effects of railway collisions confirm the experience and teaching of Sir James Paget who, speaking of the various combinations of constitutions that should be well studied in cases of nervous mimicry, says: " The most troublesome is the combination of the nervous with the gouty constitution. For in one in whom gout is not complete there are never wanting strange sensations—of tinglings, burnings, pains, pressures. . . His nervous system defines them or gives them form."* Young men make better recoveries than old, healthy men than those with pre-existent disease, and women recover not less well than men, though the hysterical disturbance may be more prolonged.

And here very appropriately arises this most important question, How far does the course of the protracted illness, apart from the nature of the original injury and shock, conduce to imperfect recovery hereafter? Remember that the symptoms have been largely those of emotional disturbance, that loss of control and feebleness of will have been at the foundation of many of them, and there can be little doubt, we think, that an unconscious or wilful yielding to every sensation that may arise, the abandonment of the conscious self to the thraldom of the morbid state, the enjoyment, so to speak of the luxury, not of woe exactly, but of gloomy hopes and feelings and fears, pave

* Op. cit., p. 195.

the way for the impossibility of regaining, even in the best of circumstances, that complete mental stability and continuous self-control which are the happy appanage of perfect bodily and mental health. A vicious habit is being impressed upon his nervous system, from which the sufferer will find it difficult in the future to rid himself. If he allows the various influences conducive to the morbid state to have the mastery over him for weeks and months, because he thinks it better to "wait and see how things turn out," unable, or making no determined effort, to resume his natural occupation and mode of living until some wholly impossible compensation has been received, depend upon it he will suffer in the future. Or worse than this, if he keeps up the morbid state by wilful means, his moral and his physical nature are subjected to a long spread-out shock from which they will find it hard to rally. As he sows so also shall he reap.

"It is, of course quite impossible for any one to avoid the circumstances which call forth emotional feeling, but it is no less certain that the development of our passions and their reaction on the bodily functions, may be very much heightened or subdued by force of habit. A man cannot alter his natural temperament, and the best resolution never to be angry, grieved, or anxious, would certainly fail, nor would it be desirable that it should be otherwise; but there is such a thing as 'giving way' to tempers and 'nursing' griefs, and fears; and in proportion as this is done, so will they become organized in our constitution, their force increased, and recurrence facilitated, until a degree of emotional disturbance is at length excited by the merest trifles which is only worthy of some great occasion."*

An interesting point in connection with these cases of nervous shock, and of the functional nervous diseases to be more specially considered in the next chapter—a point,

* Dr E. Liveing, 'On Megrim and Sick Headache,' p. 461.

however, which rather lies outside the range of our present inquiry, and has better place in a treatise on diseases of the spinal cord—is the possibility of the symptoms being due to some persistent alteration in the blood-supply of the spinal marrow. In a previous chapter we have remarked that a transient flushing of different parts and organs may perchance be the physical basis of many of the sensations which are so frequent and strange in these cases of emotional disturbance; but is there something more than this? is there, as Dr Hammond, amongst others, seems to think, any continuous anæmia or hyperæmia of the cord lying at the root of the many anomalous symptoms of functional nervous disease?

It is a practical objection to the acceptance of any such theory that neither hyperæmia nor anæmia of the cord have ever been seen, and that we never find the one or the other as a pathological condition independent of some other and more obvious lesion. It must also be borne in mind that if one part be anæmic some other part is probably hyperæmic; and difficulty must arise in precisely determining whether the symptoms be due to lack of blood here or to excess of blood there.

When we learn from Dr Hammond himself that "the principal affection with which cerebral anæmia is liable to be confounded, is cerebral congestion" ('Diseases of the Nervous System,' p. 76); and that spinal anæmia is to be diagnosticated from spinal congestion by this fact, amongst others, that in spinal anæmia there is *pain in the cord*, increased by pressure or percussion on the spinous processes of the vertebræ,* we shall not, we think, be deemed hypercritical if we say that after spinal anæmia and spinal hyperæmia or congestion the word *Query* should be writ very large.

† "In current descriptions of the symptoms of these conditions, I cannot help thinking," says Dr Gowers, "that

* Ibid., p. 417.   † Op. cit., p. 66.

a vigorous scientific imagination has contributed much more than observation has supplied." And again, in Ziemssen's 'Cyclopædia,'* we read:—"The anatomical evidence bearing upon hyperæmia in the spinal canal is as uncertain and ambiguous as possible;" and the same remark may very well apply to the opposite condition of anæmia. In summing up the evidence brought forward as to the existence of anæmia or hyperæmia of the cord, or of spinal irritation as a consequence thereof, Erb writes:—† "The conclusion could only be that we know nothing definitely at present. The most probable seems to us to be a purely functional disturbance of certain nervous elements of the cord, in company with which hyperæmia and anæmia of the cord may probably appear when the vaso-motor paths are reached by the disturbance; but this whole question seems to us awaits a solution." For our own part, while we have no evidence to offer as to the dependence of these motor and sensory disturbances upon derangement of the spinal cord itself, wherein anæmia or hyperæmia have an important share, we are much more disposed to believe that the *primary* seat of functional disturbance lies in the brain itself, and that, as in the hypnotic state induced by a profound mental impression, there is a temporary arrest in the function of that part of the sensorium which presides over and controls the movements and sensations of the periphery.‡ "Just as spasms or convulsions are more likely to happen when the

* Vol. xiii, p. 202.   † Ibid., p. 365.

‡ "The theory which ascribes the complete abeyance of the will and of sensation in hysteria to a cessation of the activity of the higher functions of the cerebral hemispheres is quite in accordance with the facts observed in this disease; if we also believe that during the absence of this governing power of the brain the spinal system is allowed to run riot, we can understand the meaning of the convulsions, strange movements, and emotional excesses so frequently witnessed in this malady."—Dr Wilks, "On Hysteria and Arrest of Cerebral Action," 'Guy's Hosp. Rep.,' vol. xxii, p. 35.

will is suspended and the cord acts independently, so when the controlling power is removed from the brain, its automatic action is intensified, and ideas exert much more power over the organic functions when directed towards them. There is a constant antagonism between voluntary and involuntary actions, and when anything occurs to neutralise the former, the latter rules the hour."*

The frequently sudden disappearance of the symptoms can only, it seems to us, find an explanation in the fact that the seat of disturbance is in the centres of conscious volition rather than in the unconscious and insentient strands of nervous fibres and cells which together make up the spinal cord.

The fundus oculi has been called the "window of the brain," and it is well known that the routine use of the ophthalmoscope affords valuable aid in the diagnosis of coarse cerebral disease. Does it also help us in the diagnosis of diseases of the spinal cord? Clinical facts are yet too few to warrant a certain answer to the question. Optic atrophy—not necessarily preceded by neuritis—occurs not unfrequently at some period or other—sometimes early, sometimes late—in the course of tabes dorsalis, and cases have been recorded in which it has also been seen in other progressive diseases of the spinal cord. With reference, however, to the diseases which have been considered in this work, we can only state that we have never been able to discover any lesion or pathological change in the fundus of the eye, except in those cases (see Cases 229 and 233, *e.g.* Appendix), so few that we make no special reference to them, where there has been actual injury to, or very near, the eyeball itself.

True it is that, as before remarked, we often hear patients complaining of their eyesight; but the failure, when real, is due either to the general prostration which may affect for the time every organ and function of the

* Tuke, 'Influence of the Mind upon the Body,' p. 99.

body, and which renders any sustained effort difficult, or much more frequently to the illness having been the means of bringing to light some anomaly of refraction hitherto unknown to, or disregarded by, the patient. We have heard the supposed changes in the fundus spoken of as congestion of the retina, whatever that may mean; but, as far as our own observation goes, too little account has been taken of the thousand and one varieties of shade and colour which the normal fundus may present in conditions of perfect health. There is, perhaps, no more untrustworthy sign, when seen alone, of pathological change in the fundus oculi than its colour; and, as has been often pointed out, more especially by Liebreich, an intimate familiarity with the varied aspects of the healthy fundus can alone fit a man to speak as to the existence of those departures therefrom which constitute real pathological change.

"The subject of the changes in the optic discs in spinal injuries," Dr Gowers writes,* "has received a large amount of attention in consequence of the prominence which 'railway cases' have given to this class of accident. In its scientific relations the subject has not escaped the sinister influence which litigation exercises on the investigation of facts, and there is no doubt that the pathological nature of many of the appearances described in these cases has been the result of an affection of the mind of the observer, rather than of the eye observed. Still it seems well established that in some cases of spinal injury ocular changes supervene, and the observations of Clifford Allbutt especially show that they occur with greater frequency the higher up the injury is. The changes are those of simple congestion, congestion with œdema, and slight neuritis, uniform redness of the disc, and concealment of the outlines so that the position of the disc may ultimately be recognised only by the convergence of the vessels. In one

* 'Medical Ophthalmoscopy,' 2nd edition, p. 169.

case a 'daffodil colour' was described. Sight is a little, but not much, affected, and the condition, which is of slow onset and course (coming on some weeks after the injury), usually passes away."*

Some interesting cases have recently been published† by Mr Bruce-Clarke ('St Bartholomew's Hospital Reports,' vol. xvi, 1880, p. 171) which show that changes in the optic fundus may be met with in cases of unquestionable injury to the spinal cord; but it is a remarkable fact that, on the view that diseases of the spinal cord and spinal membranes are so common as the result of railway accident, medical literature should contain so few cases of pathological changes in the optic fundus as a consequence of the same. And in Dr Gowers' able and exhaustive work on 'Medical Ophthalmoscopy,' one page only is devoted to ophthalmoscopic appearances in injuries to the spine.

Mr Bruce-Clarke records four cases of injury to the spinal cord in the lower cervical and upper dorsal region. In one case the injuries were very slight and no changes were observed, but in the other three there was undoubted hyperæmia with œdema of the optic disc. Of these three cases one was fatal, there having been a "fracture of the fourth, fifth, and sixth cervical vertebræ, with complete division of the spinal cord;" and in the two which recovered the optic changes were essentially *transient*. There was nothing abnormal in the pupils or in the sight, nor anything external to point to the presence of patho-

* For the ultimate issue of a railway case of eye disease the reader may refer to 'Brain,' vol. ii, p. 388.

† See also a paper by Dr Dreschfeld, of Manchester, on "Two Cases of Acute Myelitis, associated with Optic Neuritis," 'Lancet,' vol. i, 1882, p. 8, *et seq.*; and also a case recently brought before the Ophthalmological Society by Dr. Sharkey ('British Medical Journal,' vol. i, 1884, p. 1151). The conclusion seems to have been that in these cases the myelitis and neuritis were associated phenomena, due to a common cause, neither being directly dependent on the other.

logical changes. It is, therefore, possible that the same *transient* changes may have existed in numbers of instances and, because not looked for, been unknown. The author has recorded these cases, therefore, with the hope of eliciting from other observers, "whether such changes are the general rule in cases of spinal concussion at or about the lower cervical and upper dorsal regions, since further examination can alone make it clear whether such pathological changes are the general rule, or merely a rare and exceptional coincidence." If the rule, it is obvious that the optic changes will become an important diagnostic sign.*

* It may by some be thought remarkable that we make no detailed reference to Mr Wharton Jones's well-known, and in the domain of physiology most able, work on 'Failure of Sight from Railway and other injuries of the Spine and Head.' But if any one will turn to that work itself, and will then refer to the review of it which appeared in vol. xlv of the 'British and Foreign Medico-Chirurgical Review,' he will see that there are grave objections to the acceptance of the data upon which the author founds his conclusions. "Railway collisions," says the reviewer, "have been tolerably numerous during the last fifteen or twenty years, and those who suffer most are notoriously the occupants of the third-class carriages. Yet how many cases of amaurosis, caused by railway injury, appear in the out-patient rooms of our ophthalmic hospitals?"

CHAPTER VI

FUNCTIONAL OR NEUROMIMETIC DISORDERS

WE have thus been led by an almost natural step to a consideration, in the next place, of a class of cases, few comparatively in number, but in themselves most important, in that the symptoms very closely simulate those of real disease dependent on organic lesions. We refer especially to that class of cases to which Sir James Paget has so appositely applied the term neuromimesis, because the symptoms thereof are so prone to mimic those which are due to undoubted pathological change. "It is a remarkable circumstance," says Dr Wilks, "that there is no disease of the nervous system, as far as I am aware, which may prove fatal, and even show a well-marked lesion or degenerative change after death, but may have its counterpart in a functional and curable disorder."*

As has been so well laid down by Paget and others, there can be no doubt that underlying the tendency to nervous mimicry there is some predisposing idiosyncrasy or peculiarity of the nervous system of the individual. There is something about him which indicates, even to inexperienced observers, that he has a "nervous temperament;" or even if the signs of this temperament or disposition be hidden from those who meet him in ordinary life, they may be brought to light by illness or disease, and may be fully recognised by those who are called upon to treat him. No thoughtful or

* Wilks, op. cit., p. 256.

observant practitioner of medicine can fail to notice how very variable are the effects of illness upon the nervous system, and how often those who appear to show no "nervousness" while in perfect health may yet, during periods of illness, reveal that they have a "nervous temperament" which affects the course of their illness, and which must not be lost sight of if the best is to be done for the patients at the time.

It is wholly impossible, it seems to us, to define what is meant by the expression "nervous temperament." We all have probably some idea of what it means, and can call to mind examples of this peculiarity, which, under suitable conditions, may become an important clinical factor in the induction of the mimicries of disease. "In all well-marked instances" of mimicry, writes Paget, "there is some prominence and apparent excess of nervous action, leading to the general expression of the patient's being nervous or of nervous constitution. Neuromimesis cannot be found in all persons alike, or in any person at all times. It may be regarded as a localised manifestation of a certain constitution ; localised, that is, in the same meaning as we have when we speak of the local manifestation of gout or of syphilis, or of any other morbid constitution which we regard as something general or diffused, though distinct witness of it may be in only one or more parts. And the nervous constitution, like others, is inherited in different degrees of completeness or intensity ; and may, like others, become less or more complete or intense according to the conditions in which it has to live. As to what is verily the peculiarity of the nervous constitution, I believe we have nothing fit to be called knowledge. It is even hard to give fit names to what we may suppose it to be. We may speak of the nervous centres as being too alert, or too highly charged with nerve-force ; too swift in mutual influence ; or too delicately adjusted, or defectively balanced. But expressions such as these, or others that I see used, may be misguiding. It is better for us to study the nervous constitution in clinical facts."*

"Instability of functions," writes Maudsley, "is a character of the so-called nervous temperament; there is a tendency of ideas and movements to escape from the bonds of their functional relations, and to act independently—to break away from coordinate and subordinate consensus of function, and to become, so to speak, *dis*ordinate. . . . It was for this reason that I formerly described the temperament as the *neurosis spasmodica*."† "When persons have what is called a sensitive or susceptible nervous temperament, it is not merely that they are more powerfully affected in mind and body by external impressions, but that the physiological sympathy of their bodily organs is more acute and direct, whereby these answer more easily and more actively to one another's sufferings. . . . . Too close and direct a relation of dependence between the parts and the supreme authority is probably an ill thing in the bodily, as in the political, organism."‡

"Facts relating to inheritance," says Paget, "deserve great weight in the diagnosis of any doubtful case of nervous mimicry. In looking for indications of this inheritance you may not find that, in the same family, there are or have been many cases of similar mimicry of disease; but it is a fact of not less weight if, in the same family, various other forms of nervous disorders, especially of such as are, for convenience, called functional disorders, have been observed. Thus, among the relatives of those with neuromimesis it is common to find cases of mental insanity, extreme 'nervousness,' and eccentricity, stuttering, convulsive and emotional hysteria, various neuralgiæ, extremes of mental character, whether good or bad, and sometimes (but I think less frequently) epilepsy and paraplegia. These

* Op. cit., p. 180.
† Op. cit., p. 58.
‡ Ibid., p. 214.

evidences of family relations may help in diagnosis, just as in the diagnosis of a doubtful tubercular disease, it is important if, among the members of the patient's family, there have been many more than an ordinary number of cases of pulmonary tuberculosis."*

And again he writes, in a passage to which we would especially direct attention, " I believe that a large majority of the worst cases of nervous mimicry occur in members of families in which mental insanity has been frequent. And the fact is important, not only for diagnosis, but for pathology. It may serve to strengthen the view that nervous mimicry is a mental disorder; but I believe it may be more rightly read as an indication that, whatever mental insanity may be as a disorder of some portion of the brain, the like is nervous mimicry as a disorder of other nervous centres."†

We have said that it is by a step almost natural we have been led from the cases of general nervous shock to a consideration of the cases of nervous mimicry, in which functional disorders of different parts of the nervous system are very liable to arise. And the step is natural for this very reason, that the "nervous shock" has been, in many instances, the means of giving birth to that very condition of the nervous system which predisposes to the manifestation, and underlies the origin, of these functional nervous disorders. The cases, moreover, which we shall be able to quote, bear out to the full the important clinical fact which Sir James Paget—out of the wealth of his experience— has so clearly laid down, that in the worst cases we shall find some evidence, either of mental disorder in the previous history of the patient himself, or that he comes of a stock

* Op. cit., p. 192.
† Ibid., p. 193.

In his classical work on hysteria, Briquet records that in only 10 out of 396 women subject to hysterical disorders could he discover no evidence of predisposition.—Briquet, ' De l'Hystérie,' p. 396.

in which mental or emotional disturbances and peculiarities, not necessarily amounting to insanity, have been recognised as prominent in the family record. In some cases, it is true, it has been impossible to obtain any evidence of the kind, but in the absence of discoverable predisposing tendency there is adequate cause for the origin thereof in the profound nerve exhaustion, prostration, disturbance, or whatever we may like to call it, which the original shock of the accident and its varied consequences have had upon the nervous system. We need not seek further than this for a cause of the functional disorders—the paralyses, the spasms, and the convulsions—which are "mimicries of grave disease."

This view seems to be supported by the fact that some time frequently elapses before mimicry begins, time having been needful for preparation, so to say, of the nervous system for the exhibition of these disorders. But in the more serious cases, we would again repeat, there is this most powerful element also; the previous history of the patient predisposes him to serious neurotic derangement. There is no miracle in this. The higher cerebral faculties are, in such persons, less stable than in those whose history can point to an undeviating record of mental and bodily health. The higher intellectual processes are more easily put *hors de combat*, and in their temporary abeyance or annihilation lies the possibility of disorder affecting the lower centres of the cerebral or spinal mass. "Complete abeyance of the supreme functions of the nervous system is one of the most characteristic features of hysteria; so much so that where no one has yet succeeded in giving a definition of the disease, this inaction of the cerebral hemispheres leaving the spinal system to have its full play gives as good and correct an idea of its nature as any other definition. It is clear that the possessor of a very powerful will or authoritative disposition could not be hysterical except under the influence of a tremendous shock to the system, whilst on the other hand hysteria would be constantly found amongst the nervous and feeble minded."* And thus it comes to pass that purely functional disorders run in the lines of normal physiological activity ; and as the experiments made by real disease upon the brain and spinal cord give an answer to the student of disease in disturbances which flow in defined channels of morbid physiological activity, so the symptoms which are the manifestations of functional disorder run in almost precisely the same channels, and closely imitate thereby the manifestations—the symptoms, that is—of real disease. Writing, in 1866, in his first book on 'Railway and other Injuries of the Nervous System,'† Mr Erichsen said: "*Hysteria* is the disease for which I have more frequently seen concussion of the spine, followed by meningo-myelitis, mistaken, and it certainly has always appeared extraordinary to me that so great an error of diagnosis could so easily be made." Writing now, in 1882, we would here venture to point out how very often these functional disorders are mistaken for real structural disease. They are, moreover, very common after railway collisions, when the nervous system has been brought into that state which is the fit soil for their development and growth. Nor is it the least extraordinary that mistakes in diagnosis are often made, when we remember how closely the symptoms may copy those of real organic disease.

Before proceeding to support these statements by the record of individual cases, it is right that some attempt should be made to tell what we mean by functional disorders, and what are the morbid changes of the nerve centres which underlie them. That there are changes it is almost certain; what those changes are we do not know. One thing, however, may be said with some degree of con-

* 'Guy's Hosp. Reports,' vol. xxii, p. 31. Dr Wilks " On Hysteria and Arrest of Cerebral Action."
† P. 126.

fidence, that they differ very materially from the gross pathological changes in structure which we are accustomed to see upon the post-mortem table, or which, if invisible to the naked eye, are yet discoverable by high powers of the microscope. The course of the symptoms themselves, and their rapid and often very sudden disappearance, form well nigh conclusive evidence that they cannot be due to coarse pathological lesion. The change—if change there be—is in all probability analogous to those changes which are associated with each motor and sensory operation of the brain. Every volitional motor act has doubtless some change underlying it in the cerebral cortex continuous and coterminous with the act itself; and in the cortical cells there is doubtless some change which likewise underlies the thoughts of which the act is the visible expression. It is the natural change of normal and healthy function; and it is an inherent property of the cells themselves that under the stimulus of active function they should undergo such change. Nor does it do violence to physiological facts and theories to assume that under certain conditions the natural changes should become abnormal and unnatural; should, in fact, be morbid and disorderly without necessarily presenting grave structural lesion. "The conditions of nerve-tissue giving rise to functional disorder are essentially two, viz. hyper-excitability and paresis. The first may be defined as an undue readiness in the tissue-elements to assume the active state on very slight provocation, perhaps, even, without any, except what may be afforded by the circulating blood. The second is an apparently opposite state, the nerve-tissue being less excitable than normally. Feebleness and deficiency of vital action is a necessary part of paresis, but not of hyper-excitability, though it is present in the great majority of instances."*

* Handfield Jones, 'British Medical Journal,' vol. i, 1874, p. 370. "On Hyper-excitability and Paresis."
Dr Handfield Jones has many times pointed out that the very oppo-

The whole sensorium may be affected. It may pass into a state of slumber which is shown by mental hebetude, by lessened volitional power, by anæsthesia and analgesia, and, on the other hand, also by excessive activity of the automatic centres from lost or torpid cerebral control. Thus the abnormal activity of these uncontrolled centres is a symptom of the same value as that of lost volitional activity, or paralysis, due to annihilated power of the will. Diminished activity and increased activity alike point to an abnormal condition of the higher intellectual processes in the sensorium. The abnormal condition is allied in all probability to that of slumber—or at any rate this affords an explanation of the facts—and it is, moreover, akin to it in the readiness with which the symptoms may pass away when the requisite stimulus has aroused the brain from its torpid state. The needful stimulus may be some profound mental or bodily impression; it may be exerted only by the more tedious influence of re-education of the movements of the affected part; but in either case the activity

site conditions of hyper-excitability and paresis may be produced by the same causes. Thus, after recording two cases of cerebral exhaustion from pregnancy and lactation, and contrasting with them a case of cerebral excitability from the same cause, he writes :—"The exhaustion, induced by parturition and lactation, operated differently in the two cases. In one it simply impaired the production of nerve-force in the intellectual and perceptive centres; in the other it impaired especially the resisting or controlling power normally inherent in each cell, so that they were unduly excitable. In many instances these two effects are produced together. The loss of excitability in a nerve or nerve-centre, and the excess of excitability, are both alike aberrations from the normal mode of acting; in fact, signs of deterioration. I do not think this is always sufficiently recognised; the latter condition, when appearing as hyperæsthesia or hyperalgesia of some region in cases of brain or cord lesion, is apt to be regarded as an exaltation of functional power, whereas it is as truly a defect as that which declares itself by anæsthesia or paralysis, and results directly from the lesion, just as the latter does."—'Medical Times and Gazette,' vol. ii, 1878, p. 376.

of the sensorium is once more alert, and the cerebral control can be exercised in its normal and healthy way. Not less mysterious than slumber of the sensorium, whether it be of the whole or of a part, is the fact that familiarity with the morbid process seems to give the individual patient the voluntary power of putting the affected region of the sensorium into the state of torpor, or of voluntarily abandoning himself to the easily induced influence of the abnormal condition. As a wrong act once committed is more easily committed a second time, so repetition and perpetuation of the morbid vice of the sensorium make the symptoms thereof daily easier and less unnatural than they were before. The man who has once been mesmerised, whose sensorium has by the passes of the mesmerist been placed in the hypnotic state, can be more readily hypnotised again. Thus, in course of time, the "medium" of the mesmerist or of the so-called spiritualist can be reduced by the most trumpery and frivolous influences to the hypnotic or cataleptic state, and has then become the most wretched and pitiable of mortal men. "After an individual has been mesmerised repeatedly, certain movements (passes) are no longer necessary to the excitement of the sensorial volition; it has become a habit, and is produced by any insignificant associated circumstance. These phenomena may at last be thus produced at pleasure, as by an internal effort, by arresting the respiration and so arresting the circulation through the brain, &c. It is thus disinterested observers have been imposed upon; and thus hysterical girls can bring on convulsions, and any person ideas, sensations, and mental emotions, with more or less facility. Savage and superstitious nations have ever been the dupes of men and women who have discovered this power of the will over the sensorial fibres, and the brain generally. . . . . *There can be no doubt that this power of exciting real phenomena by an act of the will on the central ganglia of motion and sensation has frequently aided impostures of* *every kind.*"* The "dancing dervishes," and the "howlers" of bygone times, who lashed themselves into states of religious fury which ended in epileptic seizures, fit manifestations of "possession" not assuredly from on high, had gained by repetition of their practices the power of easy production of the frenzies which were the special attributes of their order, and which became the signs of divine enthusiasm to those who saw them writhe or who heard them scream. Beginning voluntarily, or under the influence of some wholly trivial excitement, each seizure was a sign of complete self-abandonment and loss of cerebral control, a result, it is true, of morbid action, but in no wise dependent upon pathological lesion or change. Other religions than our own may afford the best known examples of these conditions, but in this country, and in this very day, there are like manifestations of morbid activity under the influence of religious zeal.

It has been said in the last chapter that the man who voluntarily abandons himself to the morbid state submits both his moral and physical nature to a long spread-out shock from which he will find it hard to rally, and the same remark may be here reiterated with even greater force in connection with the functional nervous disorders which are results of the morbid nervous state. For there can be no doubt whatever that many of the neuromimetic conditions are more or less under the voluntary control of the patient; and, as may be especially seen in cases of convulsion, the mimetic seizures—*in themselves typical in character*—may be brought on by the will of the patient himself. And this can be done with greater ease as time goes on. Herein lies the explanation of those happily-timed convulsions which occur when it is most important that you should see them, and should be impressed by their severe reality. The seizure itself—*qua* seizure—is typical of its kind, and its phenomena lie outside the

* Laycock, op. cit., p. 111. Italics our own.

conscious control of the individual. But within his control has been the commencement of the seizure at the precise moment when it began. Thus, in speaking of the ease with which the hypnotic state may be induced in those who have been often hypnotised, Heidenhain writes : " Many of the gentlemen upon whom the above experiments have been made, need only to sit down, close their eyes, and think intently—other thoughts being excluded—that the hypnosis is coming on, in order to, as it were, voluntarily submit themselves to the charm."* And, as of the hypnotic state, so of other neuromimeses also, the patients may voluntarily submit themselves to their exhibition, and the manifestations thereof become in themselves not less real. The existence of a certain amount of control is shown moreover by the disappearance of the mimicries, when all cause for their representation is removed. The matter of compensation, as we have seen, exerts in many cases a very favourable influence on the symptoms of general nervous shock. It does so in these cases also, and examples are not few in which the typical neuromimeses came to an end shortly after settlement of claim had secured for the patients complete repose of mind, and had freed them from the necessity of any longer *allowing* themselves to be victims of the mimetic phenomena.

In speaking of the " objections made with regard to hypnotic experiments," Heidenhain seeks to show that the repetition of them does not appear to be fraught with danger or evil, but he mentions cases where attacks of convulsions constantly accompanied every hypnotic experiment, and one who suffered after every experiment from a certain degree of nervous irritability which lasted twenty-four hours.† He advises that the experiments should not in such cases be repeated, but we cannot doubt that the risk of permanent damage to the stability

* 'Animal Magnetism,' p. 86.
† Op. cit., p. 102, *et seq.*

of the nervous system and tone must be very considerable in all persons who repeatedly submit themselves, whether voluntarily or involuntarily, to the hypnotic state, even though that be not accompanied by convulsions or be not followed by "nervous irritability." The hypnotic state is not a natural, it is a morbid state ; and to repeat a vice is to perpetuate it and make it an abiding part of the organisation of the individual. Not less does risk of permanent damage to the stability of the nervous system lie in continuance of these functional mimetic disorders. The longer they exist the more prone are they to give rise to lasting functional disturbance, and to the phenomena which may be the result. And even in cases where the neuromimeses pass away under returning cerebral control, the risk is not small that from some exciting cause the conditions may be very readily reproduced. It is not unlikely, for example, that a woman who had once had functional paraplegia after a railway collision, would suffer in a similar way should she chance to be in another accident. Some cases which have come under our notice have presented the same, or very like phenomena, after two different accidents, and we can call others to mind in which there is warrant for predicting the nature of the morbid phenomena which the patients would present if it should ever be their misfortune to suffer from the effects of railway collision or severe mental shock again. The lesson to be learned from this is very obvious, that the sooner any cause for the representation of the phenomena is removed the better, and that the patients should as far as possible be freed from the hurtful sympathy of friends. There is small chance of improvement or cure so long as the patient *need not* make the necessary effort to get well, and so long as his friends, ignorant of the real nature of his malady, foster by misdirected sympathy and kindness those very symptoms whose continuance and repetition are fraught with danger to his nervous system.

The question of diagnosis is thus all-important, and once established it is doing him a grievous wrong if the sufferings and symptoms of the patient are to be made the occasion of litigation and prolonged dispute.

Two points demand brief consideration. The first is of somewhat speculative interest in connection with the central changes which may possibly underlie the mimicries of disease. Dr Buzzard has recorded a case of hysterical hemianæsthesia of the left side where there was defective electrical excitability of the right cerebral hemisphere,*

* The case is that of a girl of thirteen who had suffered a severe moral shock from the death of her father. There was at first contracture of the left arm, and when this had disappeared—after she had been a few days in the hospital—there was discovered " profound anæsthesia of the left arm, and, to a less degree, of the left half of the face and the left lower extremity." Dr Buzzard found that the interrupted current from thirty cells of a Smee's battery, which when applied to the left temple caused the girl to reel, and manifestly produced a "very powerful shock which she described as excessive giddiness," a minute or two elapsing before she recovered her equilibrium, was hardly felt when applied to the opposite side. ('Lancet,' vol. ii, 1879, p. 685.) "The day after the applications there was some return of sensibility in the left arm. The current being applied as before to the temples, no difference in the result on the two sides could be now observed."

Dr Buzzard remarks, " I cannot help thinking that the observation which I have described lends support to the opinion that the condition is a *bonâ fide* one. It will at once be recognised that, supposing the hemianæsthesia in this case to be assumed, the patient would naturally have asserted that the application on the so-described anæsthetic temple produced less shock than that upon the side in which the skin retained its natural sensibility. As I have described, the reverse obtained most unmistakably. We have it then, that in this case the right cerebral hemisphere, which, if the anæsthesia of the left half of the body were real, would be the hemisphere in fault, appeared to be considerably less excitable by the interrupted voltaic current than the left." "If supported by further observation," he adds, the fact is " likely to lend important aid towards the solution of one of the most difficult questions in neuro-pathology." Every precaution had been taken in this case to keep the child in ignorance of the application.

and the question arises whether there may not be a like condition in other functional disorders.

We have as yet barely passed the threshold of the surgical treatment of those brain lesions which, thanks to the labours of those who have experimented upon animals, we can now pretty accurately localise in the brain, but we stand also, it may be, on the threshold of another and most important method of treatment of a class of cases of a wholly different kind. In the fourth volume of 'Brain' (p. 65), Dr Althaus asks whether in the treatment of localised brain lesions it may not be expedient to utilise the "catalytic effects of the constant voltaic current in certain destructive as well as irritative lesions, directed locally to those portions of the brain where we suspect the seat of the mischief, in accordance with the teachings of pathology and with Ferrier's physiological topography of the cerebral cortex;" and he relates cases where he had galvanised the medulla, the occipital lobe, and the temporo-sphenoidal convolutions, for the relief respectively of diabetes insipidus, melancholia, and auditory delusions.

The subject so far is one of speculative interest, as yet in its earliest infancy. It has yet to be shown, we think, that good has resulted directly from the treatment, and that it is possible, moreover, to galvanise local areas of the cerebral mass to the exclusion of parts which lie beneath or around.

Secondly, and not less important to the patients themselves, is this question: Does the functional change which underlies the neuromimeses ever pass into organic disease? We can find no evidence to show that it does. "It is, I think, a fact of singular interest," says Paget, "that, in even the most turbulent of these nervous systems, the disturbance very rarely takes the form in which morbid nervous influence produces, not mimic, but real organic changes. Of the things imitated, hardly one is ever realised."* Mr Hutchinson, in one of his lectures at the Royal College of Surgeons a few years ago, threw out the suggestion, if we remember rightly, that xanthelasma palpebrarum might perchance be a result of the repeated functional disturbance which is shown by blackness under the eyes in those who suffer habitually from "sick headaches," and in whom this organic change is most commonly found. It is a suggestion only, but even if it offer a correct explanation of the origin of xanthelasma, there is as yet not a particle of evidence to show that there is any like result of continued functional disturbance in the cerebro-spinal system, even though it last for years.

The following cases afford good examples of the functional disorder which may be met with, and they will point many lessons for future guidance:

Case of functional paraplegia.—V. S—, a widow, æt. 38, the strong and healthy mother of seven children, was in a collision. There was no history of her having been much hurt at the time, but within a few hours she began to have a pain, or a sensation which she described as "opening and shutting," in the small of her back. The next morning she continued her journey of nearly 200 miles, and finding that in a few days the pain in her back was a good deal worse she went to a hospital. She was an in-patient in the hospital for six weeks, during the first three of which she was in bed, suffering from pain and stiffness in the small of the back and from general weakness. For three weeks she was up and moving about the wards, and she then made a journey of 330 miles in order to take one of her children to school. This business over she then travelled 200 miles more, home, in fact, to the place where the accident had originally happened. This was exactly two months after the accident, and as soon as she got home she at once took to her bed, suffering from great pain in the back, from much hyper-

* Op. cit., p. 183.

æsthesia in the dorsal and lumbar regions, and from general prostration. She remained almost entirely in bed until about fifteen weeks after the collision, when it was accidentally* discovered that she had lost all motion and sensation in the legs. She had complete control over both bowel and bladder, and there was neither wasting of the legs nor bed-sores. The paralysis of motion and sensation seemed absolute. The woman was at the same time exceedingly "hysterical," complained fearfully of pain in the back and of innumerable queer sensations in different parts of the body. The opinion was given that this paraplegia was not dependent upon organic disease; that it was real, not feigned; and that although there was every prospect of her recovery, it was quite impossible to say how long she might suffer from the paralysis, or how soon she might be well. She was attended throughout this illness by a trustworthy nurse, and there was never any suspicion that the woman was wilfully maintaining her condition. No material change was found in her condition up to six months after the accident when her claim—naturally a considerable one in the circumstances —was settled. Within a fortnight she left the house where she had been staying, and in three months she was walking about without assistance in perfect health, a second husband having already secured her hand and her fortune. Further account of her cannot be obtained. It is open, of course, to any one to remark that this was a case of malingering. We do not accept this view. It is of greater interest to consider what were the circumstances

* "Accidentally," because this is just what so frequently happens in hysterical affections. "It is necessary to bear in mind," Charcot says, " that hemianæsthesia is a symptom which requires to be sought for, as M. Lasèque very judiciously remarks. There are, in fact, many patients who are quite surprised when its existence is revealed to them."—Charcot, " Diseases of the Nervous System," 'New Syd. Soc.,' 1877, p. 250.

conducive to the paraplegia, and what were those which brought about her recovery. There can be no doubt that the woman received a sprain of her vertebral column and that she had some "shock," but of greater moment in the history of the case is the fact that the long and fatiguing journeys which she took within a short time of the accident must have been largely instrumental not only in preventing complete recovery from the early prostration, but even in increasing the general weakness from which she suffered. After the first journey she was compelled to go to a hospital, and after the second and longer journey she was so much exhausted that she had at once to take to her bed.

In his brilliant lectures delivered in 1881 before the Royal College of Physicians, Dr Moxon pointed out how precarious was the blood-supply to the lower part of the spinal cord. After carefully describing the anatomical distribution of blood-vessels in the cord, he says: "The tip of the cord has its blood-supply only from above, and deficiently even there, whilst the upper parts of the cord have a better sustained supply both from above and from below; and this becomes especially the case upon the cauda equina itself, for here the arteries are exceedingly minute and uncertain in size on the several nerves. Hence we see that the tip of the spinal cord corresponding to the lower limbs and sphincters is much more weakly organised as to its circulation than are the upper parts of the cord."* And he illustrates this anatomical fact by saying that "if we take the so-called urinary paraplegia, or functional paraplegia of any kind, including what is called hysterical paraplegia, we find these affections never troubling the upper limbs, but always the lower. And I think the point I have raised will suggest lines of investigation that may throw great light on this obscure class of nervous diseases.

* Croonian Lectures, "Influence of the Circulation on the Nervous System," 'Lancet,' vol. i, 1881, p. 530, *et seq.*

I believe that it is by impediment to the exceedingly and peculiarly difficult blood-supply of the caudal end of the spinal cord that all these various conditions lead to paralytic weakness of the lower limbs, and they are to be met by conditions improving the circulation if possible." The suggestion is pregnant with interest, and herein it seems may also be the explanation of that real weakness which is often complained of in the legs in cases of general nervous shock. In the patient whose case we have recorded, each event after the accident must have tended to reduce the strength of the circulation, and so directly to affect the nutrition of the lower part of the spinal cord. We do not, however, feel satisfied that this can be the explanation of every case of functional paraplegia, and we are inclined to think that it must be looked for in psychical rather than in physical causes. It is a noteworthy fact that in a large majority of the cases of functional paraplegia there is little or no paralysis of the bowel and bladder, and it seems strange that if paralysis of the legs be due to malnutrition of the spinal cord, the bowel and bladder should not also be affected. Functional paraplegia is characterised, moreover, by being *absolute*. The patient seems wholly unable to move the legs or to feel in the very least degree, seems to lack all power of bringing the influence of the will to bear upon the lower extremities. " In true paralysis the spinal system is affected, whilst the will is good, hence the patient is seen to make the greatest effort to move a leg or arm, although the result may be ineffectual. In disease of the brain proper, or during its functional abeyance in hysteria, it is the will itself which fails."* And although we feel hesitation in offering any explanation of these cases after the suggestions of so competent an observer as Dr Moxon, we cannot help thinking that in many of them the determining cause of the paraplegia lies rather in the sensorium than in the

* Wilks, 'Diseases of the Nervous System,' p. 114.

spinal cord. "They are rather functional diseases of that part of the brain which has to do with the movement or sensations of the limbs, than functional diseases of the cord."* It is because the seat of the paralysis is not in the cord that the bowel and bladder are so rarely affected. Desired action of these organs is told to the brain by repeated conscious impressions which are absent from movements of the legs, movements, moreover, which need not necessarily be made. There is probably enough volition also to prevent the unpleasantness or discomfort which are inseparable from paralysis of bowel and bladder.

The mode of recovery too seems to show that the morbid focus must be placed in the brain. A sudden impression compelling the automatic use of the legs may in a moment arouse the torpid sensorium, or, as we have seen in the cases of strong and healthy girls—hospital patients, with no sign of feeble circulation—a process of re-education, beginning at the very beginning, as it were, may be needful to restore the lost movements of the limbs by reawakening the brain to a full sense of its responsibilities in the circle of the will. Cases like these may look like fraud, but we feel sure that in many of them we must agree with Paget that the "fault is rather in weakness of the will than in its perverse strength."† The patient says, "as all such patients do, 'I cannot;' it looks like 'I will not,' but it is 'I cannot will.'"‡

The case of functional paraplegia which we have recorded was a very typical one of its kind, but we shall do well to draw our remaining examples from patients of the sterner and usually less hysterical sex.

Case of functional motor paraplegia. Extreme emotional disturbance.—T. B—, æt. 41, a man of gouty family, naturally very excitable, and able, as he said, to hear a pin drop in the next room, was in a very severe collision in

---
* Gowers, 'Diseases of the Spinal Cord,' p. 65.
† Op. cit., p. 189. See also Case 97, Appendix. ‡ Ibid., p. 188.

which the carriage he was in was smashed to pieces. He crawled out of the *débris* as best he could and went on his journey, but in about half an hour he began to have retching, pains in the abdomen, and shivering. He therefore returned home. There were slight bruises about the limbs, and one on the forehead; and the next day, when in bed, he complained of pain in the right side of the abdomen and the lower part of the back, but at neither of these places was there mark or tenderness. For the next few days he seemed very ill, had severe pain in the head, occasional retching, and at night he wandered. For three days his temperature was raised. He was in a highly nervous state, and spoke frequently of a dread of lock-jaw and paralysis. Three weeks after the accident he still complained of severe pain about the sacral region, but there was no tenderness. He complained also of "numbness" in his legs, a word used by him to express not impaired sensation, but a difficulty which he felt in moving them. There was no hyper- or an-æsthesia, but his walking, in which he helped himself by holding on to the furniture, was done with apparent fear and effort. He could stand quite well with his eyes shut, and there was no spasm of the muscles of the legs. His temperature and pulse were normal and the bodily functions were naturally performed. His general condition improved, he was able to eat and sleep better, and even to get out of doors in a chair. He still suffered, however, from extraordinary emotional disturbance, was very irascible, and frequently cried. He continued to dwell on the fear of paralysis, and steadily lost the power of moving his legs. He made for himself an ingenious contrivance whereby he was able to move about by the support of his arms, but his legs were hardly used at all in progression. Eight months after the accident he was quite unable to walk, and failed entirely to make any requested movements of the legs or feet during examination. The attempt to move his legs produced great mental agitation. There was no paralysis of bowel or bladder, and sensation of the legs was but very slightly, if at all, impaired. There was no material wasting. The cremasteric reflex* was normal. There was no reflex spasm, and no sign of bed-sore.

Nine months after the accident he had an attack of aphonia, brought on suddenly by hearing of the death of a friend. The aphonia lasted for three weeks and then disappeared as suddenly as it began, when startled by one of his children rushing into the room. He also suffered from frequent nausea and retching, the least excitement, such as the visit of a friend, almost certainly making him sick. It is indeed very difficult to express in words how extreme was the emotional or hysterical disturbance in this patient. He had always been a man of highly nervous temperament, likely, so it was said, to suffer severely from the shock of a railway accident.

We failed to satisfy ourselves that the paralysis was dependent on organic lesion, and eleven months after the accident we reported to the railway company that the "cause of the paralysis seems to lie rather in the directing power of the will than in lesion discoverable of the brain or spinal cord." We further expressed the opinion that the case was perfectly genuine, that it was wholly impossible to say how long he might be ill, and, moreover that litigation would be exceedingly detrimental to him. To the man himself we advised his making every effort to use his legs, and to re-educate the movements of them by daily practice. Litigation was avoided and the claim, naturally and rightly a very large one, was amicably settled thirteen months after the accident. By the kindness of his medical attendant we had frequent reports of this man after his claim was settled. For long he did nothing and remained in a nervous hysterical state, and it was not until four

* This case was seen before the value of patellar and other reflexes was known—and the cremasteric alone was tested.

years after the accident, when he made a complete change in his living and occupation, that he began to get well. We saw him seven years after the accident—in the course of the present year, 1882. His own story will speak better of his condition, past and present, than any other words. He considers that he was ill for between four and five years. He used the appliances for walking for about two years, and then began to use sticks. Two years and a half ago he took a public house in the country, and began to lead an out-door active life. When he first began this, he could not get up from his chair without help, having always to be helped up, or to pull himself up by getting hold of something in front of him. Suddenly one day he got up without knowing it, and his son said to him, "Why, father, look what you've done!" "Good God!" he replied, "I have got up myself." From that day forth he was able to get up without difficulty. He still has great fear about his spine, and only a few weeks ago when a friend struck him in the back in joke, he was terribly alarmed, and for two days could hardly walk. He can walk nine miles without fatigue and ride all day, he has gained weight, and is altogether stronger and better than he was before, regarding his recovery as due to change of life and scene. In appearance he is the picture of health, and as far as his legs are concerned—and they were very carefully examined —there is not a sign or symptom of anything whatever amiss with them.*

Case of supposed spinal injury. Functional spasmodic twitchings of arm, &c.—S. B—, æt. 33, was in a railway collision at night when a large number of persons were more or less shaken and hurt. He himself was not injured, as far as he knew, at any one place, and no marks of external injury were at any time discoverable. He com-

---

* See a remarkable case recorded by Dr Webber ('Boston Medical and Surgical Journal,' vol. x, p. 44, 1872), "Recovery after four years' paralysis following railroad injury."

plained, however, of being shaken, and looked pale and ill. He took to his bed, and in a few days complained very much of his back, and was in a continued state of alarm about his "spine." Beyond appearing shaken and nervous about himself, he had no sign of structural injury to any one part of the body. He remained in this negative condition for some weeks, and then began to move about the house, and once or twice he went out of doors. About this time there came on a peculiar twitching in the left arm, which is thus recorded in the notes:—"To-day on our arrival he was lying dressed on his bed. We asked him to go into the next room, and he got up without apparent difficulty and did so. He sat down in an easy chair, when his left arm and hand at once began to jerk with sharp clonic spasms or twitchings, not unlike the movements of chorea. The movement kept on when his arm was held, and he said he could not control or arrest it. It was noticeable, however, that it ceased entirely when he began to undress, partially ceased when he engaged in conversation, and altogether stopped when his attention was specially directed to some other part of his body. Coincident with this movement of the arm was a continuous jerking of the head. There was no wasting nor any sign of loss of power in the limbs." He complained greatly of his back, and evinced tenderness on touch at the mid-dorsal and upper sacral regions. The temperature was normal, and all the bodily functions were naturally performed. He continued in much the same condition for nearly a year, a severe injury to the "spine" being made the basis of a demand for large pecuniary compensation. There was, however, neither history nor sign of lesion in any central structure, and the whole condition was regarded as one of functional disturbance which might be very much controlled if the patient would only choose to exercise his will. This view of the nature of the case received strong support from his previous history. He had been in a railway collision twelve years before. He then received no bodily injury, but he was very nervous about himself, and four months afterwards began to suffer from spasmodic wry-neck, which lasted for four months, and which recurred again for a short time after an interval of two years.

There was no reason in this case to attribute any want of *bona fides* to the man in the presentation of his symptoms, although the largeness of his claim and the sequel of his case would rather tend to throw doubt upon its entire genuineness. When compensation was settled he very speedily lost all the spasms and returned to work, and it was even said by one who had taken a friendly interest in his case that he had "recovered with indecent haste." But his recovery was due, we think, rather to the fact that settlement of his claim enabled him to make the requisite effort to do something, and that healthy occupation provided the means of diverting his attention from himself and his ailments, so that the spasms were unconsciously forgotten and forthwith disappeared. Five years after the accident the report ran that he was "in good health, though he had been shaky and nervous" for some considerable time after his claim was settled.

Both these cases present us with examples of undoubted predisposition to neurotic disturbance. It is wholly impossible to say why the functional disorders should have assumed the forms they did, but it is interesting to note that in both there was a genuine dread of spinal injury, and that in the second case the wry-neck after the former accident, and the chorea-like movements of the arm and head after an interval of no less than twelve years, were disturbances of the same kind. The cases, moreover, show of how much importance it is towards a correct diagnosis to know something of the previous history of the patients, and to their special *liabilities* of disease.

Shock to nervous system. Hysterical seizures.—R. C—, æt. 39, an officer in the army, was in rather a severe collision at night. He was awake at the time and was thrown backwards and forwards in the carriage. He had no knowledge of being hurt, and helped the stoker who was much injured. He then finished his journey, the "excitement," as he supposed, "keeping him up." The next morning he felt very ill and vomited, and he soon began to suffer from pain across the loins, queer sensations all over the body, nausea, giddiness, and want of sleep. On the third day he took a long journey of several hundred miles to be with some friends; and on the twelfth day after the accident he suddenly fell and struck his nose against the corner of a table. He soon became conscious and screamed violently. His own description was that the "fit came on about three in the afternoon, I fell down and screamed, and then began to cry and sob violently. During it I was unconscious, although I knew that people were around me, and that I must use all my efforts to restrain myself and to keep quiet. When all was over, I did not know what had happened." He called this fit an "hysterical attack," and the doctor who saw him immediately afterwards, and who found him more or less unconscious, thought that this was its nature. Six weeks after the accident he complained of pain in the back, loss of memory, inability to apply himself, occasional giddiness, nausea, and want of sleep. He looked anxious and worn, and his doctor, who had known him for some time, said that he was undoubtedly much changed in manner and appearance. He had lost flesh, but all the bodily functions were natural. He described the fit in the words which have been given, and said that he had had two or three since, though not so violent as the first. Shortly after having left the house, we were summoned to see him. We learned that an attack or "fit" had begun with screaming, and we found him lying on the sofa with his eyes closed, his face very pale, and his pulse small. He took no notice of our entry into the room, but occasionally sighed. Asked how he was, he opened his eyes and looked wildly about. He was then very sick. After vomiting he roused himself, asked how long we had been there, and said he was better. Before we left he was apparently asleep. Within the next two months he had three or four attacks of the same kind, though of gradually lessening severity. His claim was settled six months after the accident. Twelve months afterwards he still suffered in a "slight degree" from the effects of the injury, but it did not prevent him from attending to his duties. No later record than this can be obtained, that six years after the accident he was still on "active duty." A feature which cannot be omitted from the history is that this patient wrote repeated, and what might almost be called "hysterical," letters about his condition and future prospects, and that he made an enormous claim, thrice the amount which he ultimately received, without resort to litigation (see also Cases 1, 20, 59, 149, 230, Appendix).

Nothing is known of this patient's previous history, other than that he had always enjoyed good health; and the question arises, what was the origin of these hysterical seizures? It seems to us that in all probability they began with syncope, which was a direct result of weakened power of the heart from the nervous shock, and that they assumed the form they did from the very fact that the accident had produced a profound impression on the patient's mind. He was reduced to a condition in which he was ready to be alarmed, and when, after the fainting, he became partially conscious upon the floor, he screamed hysterically in very natural and increased fear. And each subsequent fit began in the same way, by a sensation of syncope—not perhaps amounting to actual fainting—which by the alarm it caused him at once determined the screaming and sobbing which were characteristic signs of each attack. With returning strength and cardiac tone the seizures lessened in frequency and severity, until at length they died away. Had the patient been a woman the fits might have lasted for a much longer time, and been frequently induced by altogether trifling causes, causes wholly different perhaps from those which originated the first attack.

We shall proceed in the next place to the consideration of cases of exceeding rarity and interest for which it is most difficult to find any appropriate name. They may, however, very rightly be placed in that class which we have called functional disorder of the nervous system. The psychical disturbances are more profound than in any of the cases yet recorded, and in the account of them it will be well to bear in mind what has been previously said as to injuries of the back. The psychical disturbances are indeed so great that when they are accompanied by some slight injury to the vertebral column, they seem almost to shut out the possibility of injury to the spine or its contents being the real cause of the symptoms. And in the consideration of injuries to the spine or its varied structures and contents without apparent mechanical lesion, cases of this kind are of especial value, in that they are merely extreme examples of those far commoner cases, where there is a lesser degree of mental disturbance and prostration, and in so much a greater risk that the injury to the back may be looked upon as the chief element in the case, and the treatment applied thereto become a source of much evil to the patient.

In the two cases which are now to follow there was a history of previous psychical disturbance in the patients themselves, and of mental instability or insanity in their families. It is very clear that both were predisposed to suffer from neurotic disorders should any adequate cause for them arise.

Case of hypnotic catalepsy, &c.—B. A. B.—, æt. 36, a strong and active man, was in a railway collision at night, in which a large number of persons were more or less injured, though the accident was not severe. He complained shortly afterwards of having been shaken, and also that his back had received a wrench, owing, he thought, to his sitting sideways when the collision occurred. He had one or two slight bruises on one arm and a sprain of one wrist. For the first few weeks after the accident there were no symptoms of constitutional disturbance or of serious injury, but the man said that he could not hold himself upright, or walk any distance in consequence of the injury to his back, and the doctors who saw him thought that he was to some extent exaggerating the effects of his injuries. About five weeks after the accident he suddenly changed; he constantly repeated that he was going mad, and that he was sure he was going to be paralysed. He began at the same time to take violent exercise, walking several miles a day at great speed. This was followed by a condition of exhaustion, in which he was "wandering and hysterical," and which was described by a medical man who saw him as "hysterical mania." This lasted for several days. There succeeded to this a state which can only be described in the words recorded at the time. "He is lying in bed on his right side with his knees drawn up. There is not the slightest movement when he is spoken to, or when he is touched through the bedclothes. There is a continuous quivering of the upper eyelids. Asked to put out his tongue there is no response, though when the lips were pulled apart he seemed to make some effort to open the jaws and protrude the tip. By raising the lids the pupils are seen to be equal in size, and they react normally to light. The aspect of his face is that of complete repose and disregard, but he is obviously not entirely unconscious. Pulse 56. His arms and hands remain in any position in which they are placed. The arms and legs are very much wasted, and the whole body seems emaciated. The legs are at once drawn up spasmodically on tickling the soles, and pinching the calves evidently causes pain, for he groaned and much contorted his face. On touching any part of the chest or abdomen rather firmly with the fingers, the whole body, face, and arms are spasmodically worked, the legs being frequently abducted and adducted. The abdominal muscles are almost as hard as a board." He is said to have occasionally an "hysterical fit," consisting of spasms all over the body, beginning with an expression of fright, and lasting about fifteen minutes. An experienced nurse attending him says they are not like epileptic fits. He takes plenty of nourishment, milk and beef tea, but little or no alcohol. He passes water only once within twenty-four hours, sometimes groaning beforehand as if in sign to the nurse. The bowels are never moved without enema. He lies for hours absolutely motionless, and three weeks ago he never moved a finger for a whole day, nor passed water once. A serious feature in the case is the great wasting, food although taken in abundance seeming to have small influence in maintaining the bodily nutrition, and he looks as if he might sink and die. This condition lasted for about six weeks, and then under the influence apparently of larger doses of alcohol—the increasing exhaustion and wasting having seemed imperatively to call for it—he began to emerge from the state in which he was, to move in bed, to open his eyes, to take more solid food, and even to speak a little very feebly. He was soon able to get up and go about, made flesh again rapidly, and took some exercise. He was, however, very nervous and apprehensive, and felt sure he should never get well. Seven months after the accident he still complained of his back, and held himself in a stooping posture. Questions were answered very slowly, and any required act, such as that of putting out the tongue, seemed to demand an unnatural effort. From this time he continued to improve, and in eight months he was so far well that it was thought right and prudent to allow him to arrange his compensation.

It bears upon the case that the claim was by no means large, and there was no reason at any time to believe that the matter of compensation was unduly affecting the patient's mind. Of far greater importance is the fact that there was a strong family history of insanity. His father and one uncle were "queer," a brother had actually been in an asylum, and his sister is very hysterical. His own account of the condition in which he lay so long is "that he knew all that was going on around him, that he remembered when the doctors came, and knew always when there were more of them than usual, but that he could not speak, and supposed that his brain would not direct him to do so." The sequel of the case is satisfactory, the following report of him being obtained two years after his claim was settled, or thirty-three months after the accident:—"His recovery was gradual, but without any relapses. He married six months after his claim was settled, and has one son about two months old. He has had no illnesses, is at present strong and stout, and is emigrating some time this month."

It is impossible to conceive that the symptoms in a case like this could have been in any way dependent upon injury to the spine, and this much may be said of it that, when this strange condition supervened, all thought of injury to the spine as a cause thereof passed from the minds of those who were attending him. The man, indeed, had himself shown by the violent bodily exercise which ushered in the mental disturbance that there really was no sign of paralysis or even weakness in the legs, and that the pain in the back was very slight indeed. The condition was essentially one of profound mental disturbance originated by shock, immediate fright, and the fear of impending evil, in a man with a strong family taint of insanity. The higher cerebral faculties seemed for the time to be in that state of slumber of which we have already spoken, and the general state was very like that which has been described by Heidenhain* and others as occurring in the so-called mesmeric or hypnotic state, and associated with cataleptic phenomena.

Almost exactly parallel with this case is that of a strong and healthy man, æt. 30, who was in a collision, and who presented the usual signs of having received a sprain of his back and some general shock to his nervous system. He lay for long in much the same hypnotic state as the last patient, alternating with fits of violence and passion. When he awoke from this, he became the subject of a delusion that he was being poisoned, and was accordingly, about ten months after the accident, removed to an asylum. He remained there about six weeks; and while an inmate he adopted a peculiar gait, which lasted up to the time when his claim was settled two years after the accident, and which formed the ground of a very serious view that he had received a permanent damage to his spinal cord. His mode of walking was thus described when he came out of the asylum: " He puts the weight of his body on two sticks placed in advance of him, and draws each leg alternately forward with the foot much everted. When about to advance one leg he twists the other inwards on the toes, so that the latter point forwards instead of outwards. He keeps the knees quite stiff. In this way he shuffles along with great rapidity. As he stood with his back against the wall, he was asked to lift up his knee, but

---

* Writing of the disturbances of the motor apparatus which have been observed during hypnosis he says: "More or less extensive cataleptic rigor becomes established; the limbs thus affected remain in any imaginable position they are placed in. The will has, it is true, not wholly lost influence over them, but it is exerted with very great difficulty. If, however, with a great effort, the parts be set in activity, there often results, instead of simple, convulsive movements which spread to other parts of the body."—Op. cit., p. 77.

he professed utter inability to do so." Very careful examination was made at this time as to the nutrition and state of the legs, and a report shortly afterwards by a very able physician runs thus : " The reflex irritability and faradic excitability of the muscles of the lower extremities are normal; there is an entire absence of affection of the bladder or rectum, or of any trophic change, such as muscular atrophy and bed-sores. There is also an entire absence of muscular tension, rigidity, contraction, or deformity, in the lower limbs. Examination did not enable me to determine whether any affection existed on the sensory side, as the patient absolutely refused to answer any questions. On the whole, my opinion of the case is that it is an example of many recorded instances in which a slight and unimportant injury develops various emotional and hysterical symptoms." At a final visit made to him before his claim was settled he complained more than ever of pain in his back, and called out loudly when touched upon his clothes. While sitting in his chair he could move his legs in any direction required of him, though much persuasion was necessary to get him to move them at all. He suddenly vomited during our visit, without any precedent sign of nausea or retching. Asked to walk across the room he essayed to do so after much urging, and walked in the manner already described. There was no tremor of the legs during progression, and nothing like ankle-clonus or the gait which is seen when there is secondary degeneration of the cord. Subsequently, on being asked to go into the next room he began to do so, but almost immediately fell down flat on the floor, whence he was lifted and carried away. A very large claim for compensation was preferred, and was arranged two years after the accident, not, however, without a resort to litigation. He shortly afterwards left the house in which he had been living, and for some time it was not known where he was. Forty-two months, however, after the accident he was fortunately seen by one of the medical men who had visited him during his long illness, and he found him in perfect bodily health and vigour and the father of another child. It should be stated, as having an important bearing on the case, that the man's previous history was bad. He was always very irascible, and some years previous to the accident he had been laid up with sunstroke. There was also some doubtful history of insanity in his family.

It will not be thought that this case has been mentioned unnecessarily when we point out that pain in the back was throughout a prominent symptom, and that it was considered by some to be a case of severe injury to the spine. From the first moment, indeed, treatment was specially directed to his vertebral column, and a most careless examination of the urine, which was found to be feebly alkaline after it had been standing for some time, seemed to lend support to the diagnosis that there had been injury to the spinal cord.

It need cause no surprise that there were wide differences of opinion as to the nature of this case. "Shamming," on the one hand, to sclerosis of the lateral columns, preceded and originated by a meningitis, on the other, formed the two extremes. The truth lay between them, and that opinion turned out correct which held that it was essentially a case of functional disturbance, and that as there was no special reason or symptom to place any lesion in the spinal cord, the man would in all probability get perfectly well. The previous history of the patient showed that he was liable to serious psychical disturbance, but it is only right to add that the motive in this case for maintaining the neurotic state was exceedingly strong. We believe that control might have been exercised by this man far more easily than by the patient whose history came before (vide *ante*, p. 238). The diagnosis of "shamming" was not ours, but from various circumstances of the case, to which we need not here refer, we hold that he approached much nearer wilful representation of the symptoms. That there never was any lesion of the spinal cord the issue of the case has abundantly proved, and we cannot help thinking that such a diagnosis would never have been raised were the influences of the mind upon the body more fully recognised, and were it not unfortunately regarded as almost a matter of course that the injuries received in, and the symptoms seen after, railway collision must be due to " concussion of the spine," and be followed by the chronic meningitis, and the myelitis, and the " inflammatory irritation of the membranes," of which we hear so much but which no man has yet seen.

CHAPTER VII

MALINGERING

Is the condition before us real or feigned? is a question which most of us have sometimes to ask ourselves in the routine of daily practice. A right answer is obviously fraught with moment to both doctor and patient, and yet the difficulty of giving a right answer may be very great. The simulations of disease are so many and various, aberrations from typical symptoms of disease are so unaccountable and strange, the idiosyncrasies of individuals, and the motives whereby they may be influenced, are so obscure, that the diagnosis of a feigned disorder may become well-nigh impossible. That a study of feigned diseases, however, is not without interest and importance, let the special treatises devoted to them in many languages, and let the chapters thereon in any of the standard works on medical jurisprudence bear witness. We do not here intend to deal with that kind of malingering about which much has been written, nor in briefly directing attention to the study of feigned diseases, to draw our observations and conclusions from sources which are as open to others as to ourselves. We desire rather to offer some remarks on the special kind of malingering which is to be met with after injuries, or after no injuries, received in railway accidents; for although it has not been our purpose in this work to treat especially of the darker side of human nature, it is nevertheless a fact that there is no class of cases which in the present day so frequently suggests these questions to the practitioner of medicine :—Is the condition before him real or feigned? is the story told him false or true?

It is requisite, however, in the first place, to say something about malingering in general, and of those conditions more especially which, without due care, we might regard as fictitious when they are really genuine. Turning to a recent and able work on 'Medical Jurisprudence,' by Prof. Ogston, we find the following classification of feigned diseases:*—" (1) Feigned diseases, strictly so called, or those which are altogether fictitious. (2) Factitious diseases, or those which are wholly produced by the patient, or at least with his connivance; and to these have been added by some writers, (3) Exaggerated diseases, or those which, existing in some degree or form, are pretended by the party to exist in a greater degree or different form; and (4) Aggravated diseases, or those which, originating in the first instance without the person's concurrence, are intentionally increased by artificial means." It will serve no practical purpose to adopt this classification as our guide, admirable though it be; for we shall find clinically that cases often fall into one category quite as well as into another. It helps, however, to some extent as a pioneer and to clear the ground.

A large number of the cases of real malingering of which we read in books are drawn from certain sections of the community. The soldier—proverbial when "old"—the sailor, and the prisoner in jail have provided many of the most striking and remarkable of published cases. In such there are obvious reasons for assuming disease when they have it not; the soldier or sailor to avoid foreign service, or the dangers of battle; the prisoner to escape the daily toil which is his penalty and his portion. It is a common characteristic of most feigned diseases

* Lect. XXIII, 'Lectures on Medical Jurisprudence,' 8vo, 1878, p. 334.

that they have in them some trace of reality, some foundation, so to say, of the symptoms complained of, or of the physical condition shown. The prisoner who declares that stiffness of his knee prevents his working on the treadmill, or stiffness of arm and back the turning of the crank, has revived perhaps the memory of an old bruise or pain which he now says gives him trouble. That is our own experience and others have said the same. The duty of the surgeon, nay, the very essence of his calling, is to be able to recognise and estimate the value of the state before him, or of the story which he has been told.* Depend upon it, if a man has not known disease at the bedside, if from want of familiarity with disease he cannot rightly weigh and balance its different symptoms and signs, he will be almost certainly deceived when a case of fictitious disease comes before him. An apt though simple illustration of the risk of deception may be taken from the attitude and position of joints in real and assumed disease. He must be more expert than are most malingerers who can make his knee-joint, for instance, show the ordinary symptoms of veritable disease of synovial membrane, cartilage, or bone. Let us put aside the physical signs—heat, swelling, and the like, which may be absent—and glance only at the position of the limb. "Permit me," writes Mr Hilton in his lectures on 'Rest and Pain,'†"to refer to this constantly flexed state of an inflamed joint: Take, for example, that of the hip: I venture to say that no gentleman here ever saw an inflamed hip-joint with the leg extended. . . . In the case of the knee-joint, when inflamed it is always flexed. Curiously enough, the malingerer, wishing to

---

* "It is the doctor's daily labour to unravel the meaning of pain, whether it has a real seat or whether it is subjective. No rules for diagnosis can be laid down; every case must stand on its merits."—Wilks, op. cit., p. 374.

† 2nd edit., p. 156.

deceive and to impose, almost always endeavours to indicate his long-continued and extreme suffering by fully extending the leg. But this extended position displays the imposition."

We may meet, moreover, with patients who, while not pretending to be afflicted with actual joint disease, nevertheless affirm to you that joints have become stiff, and that they cannot move their limbs. Reference need not be made to the well-known and recognised causes of stiffness and anchylosis of joints, suffice it that the entire absence of them, whether it be in the history of the case, of the injury inflicted, or the course of the disease, will reveal a flaw in the evidence sufficient to raise suspicion. Suspicion grows into tolerable certainty if we be careful to observe the conduct of the person under examination. A man came to the out-patient room of St Mary's Hospital complaining that he could not work because of stiffness in the right elbow and inability to straighten his arm. He said he had fallen on his elbow a month before, but it was clear from his statements on questioning that the injury had not been at all severe. Comparison of the two elbow-joints showed an entire absence of physical signs, and there was no wasting of the limb. Noticing, in examination, that attempts to flex or extend his arm were forcibly resisted, he was told to look in the opposite direction, questions were asked him unconnected with his arm, and there was at once no difficulty in bringing it to natural and full extension. A turn of his head and eyes towards the affected limb was immediately followed by active flexion to the original degree. Complete flexion also could be produced under like circumstances. Examination of both arms simultaneously seemed to confuse him, for he called out with pain when pressure was made on the sound limb. Such inconsistencies as were met with here ought, at any rate, to place us on our guard, and we should hardly have thought it necessary to point them out, did not we sometimes meet with the like methods of imposture, and had we not known of cases where the deception had been successful. A man based a large demand for compensation from a railway company on stiffness of his elbow and inability to move his arm, the result of a collision. A verdict incommensurate with his expectations having been recorded, he threw up his arms and exclaimed, "My God! I'm a ruined man." With greater decorum both of language and gesture, he might have waited until the rising of the Court before he moved his rigid limb.

Take another view of the impostor's art. We may find him adopting devices to produce conditions which, in themselves alarming, are yet seen to be without significance when every feature of the case is examined. A prisoner took to his bed complaining of great pain and swelling of the abdomen. Although the belly was enormously distended and tympanitic, there was no other sign of illness about him, and there was an entire absence of any one condition on which tympanitis usually depends. After a few days' observation, and having carefully weighed all the facts of the case, the surgeon came to the conclusion that the man purposely induced the distension by swallowing air. Loudly enough for the prisoner to hear him, he accordingly whispered to the warder, "When I come tomorrow I shall bring an instrument to tap him." On the morrow the tympanitis had disappeared.*

A further illustration may be drawn from cases—by no means uncommon—where patients assure you that they are losing blood in large quantities from the bowel. Now

---

* The same method of imposture is recorded by Gavin ('Feigned and Factitious Diseases,' p. 299), who writes: "This affection, tympanitis, has been so successfully feigned as to deceive a board of French medical officers; but this individual possessed the extraordinary power of greatly distending his abdomen by swallowing air. He, however, obtained an unqualified exemption from military service by presenting himself in this state, with clothes made for the occasion."

the causes of hæmorrhage from the bowel are well known, though we must admit that, from their very number and variety, an exact diagnosis may be exceedingly difficult. The malingerer, however, forgets that profuse hæmorrhage —and it is of profuse hæmorrhage of which he invariably complains—gives rise to well-defined symptoms due to loss of blood. Who ever saw a patient losing blood, either in alarming quantities or in small amounts spread over a long time, with a florid lip, a tranquil pulse, a cool skin ? Should not the presence of every indication of health warn us that we have to deal with something altogether unusual ? Should we be doing rightly if we paid little or no attention to the general condition of the patient, and endeavoured to estimate his case by simply hearing the story he told us, without examining the blood which he showed ? Thus it was that an erroneous diagnosis was made of a case where pints of blood, not the man's own, were presented as having been passed *per rectum;* thus also in another case where a man showed from the same supposed source prodigious quantities of " blood and corruption." Both were cases of imposition —subsequently known and proved—after railway accidents; and yet in both the fraud was successful, because there was neglect of the precautions we have named. Given such conditions as these absolutely alone, without cause discoverable, or without result, can we not say of them at once, " Impossible, untrue ?" Slight tingeing of hardened fæces with blood after the bowels have been constipated from a few days' confinement in bed, gives the cue perhaps to the man who is ready to utilise and magnify anything abnormal which he can find ; and careful examination in the early days of such complaints would do much to arrest the development of outrageous fraud.

In the same category it is not amiss to refer to dilatation of pupil induced by the use of atropine. " The access to atropine or belladonna," writes Mr Hutchinson, " on the part of the public is now so easy that we cannot be surprised that we encounter mydriasis as the result of an accidental and perhaps unknown use of this agent, or of its use with intention to deceive. It is the first question which will occur to a surgeon on seeing a dilated pupil, 'Has atropine been used?' and he must be on his guard in cases of hasty denial. Many patients deny at first that they have used drops, in whom cross-examination will establish the fact. Either it had been forgotten, or the drops had been put in by a chemist or surgeon, and not thought to be of importance. Not unfrequently, however, the ophthalmic surgeon has to encounter cases of intentional deception. These occur usually in young women of emotional tendencies. Not a week ago a highly cultivated young lady consulted me for 'pemphigus.' She had blebs all over the left half of her body. But these blebs were, some of them, not round but oblong, in a style which no skin disease ever assumes, and very obviously the result of the application of a brush. She was liable also, I was told, to attacks of dilatation of the pupils and loss of ability to read. These attacks usually lasted a week. This case is only an example of what has frequently come under my notice. Although it is possible to use atropine in such a weak solution that the ciliary muscle is not affected, yet in most of these cases a more complete effect is obtained, and the loss of power to read is produced in addition to mydriasis. If the latter be present alone, and if it persist for long, the suspicion of deception may be put aside."*

We shall not attempt to enter in any detail into the feignings of paralysis and kindred nerve diseases. The same principles must guide us, and we shall find them very seldom fail. The artifice may be clever, and well devised; it may be long sustained and free from variation; but it is rare, most rare, most difficult for the malingerer to simulate with accuracy a real disease. He exaggerates;

* "On the Symptom-siguificance of different states of the Pupil," 'Brain,' vol. i, p. 462.

that which he could not do he does; he will not do that which he could do if his state were real; and you find that his symptoms are such as you have never seen or known resulting from any affection of the brain, the spinal cord, or the nervous system generally. Subjective symptoms largely predominate, and we shall observe that such objective symptoms as he has are mostly those over which he can exercise his will. He cannot make his eyelid droop; his tongue does not always deviate, nor is the angle of his mouth always drawn; he knows not how to paralyse his bladder; there are no bed-sores; he does not waste; his palsied limb resists examination; his fits occur at convenient moments when he cannot harm himself, or when he can be under the observation of those who know not their import or their signs; in his coma he is not unconscious, and added to all he has the aspect of health, nor is any vital function deranged (see Case 181, Appendix).*

Thus much for some of the grosser and more obvious simulations of disease. It behoves us in the next place to consider the opportunities which old injuries, previous disease or change, may give the malingerer for the practice of imposture. Hydroceles, varicoceles, herniæ, fatty tumours, sloughing gummata, sebaceous cysts, and distended bursæ are some of the conditions which we ourselves have seen people attempt to palm off as having been caused by the shock of a railway collision. Surely such patients must fondly imagine that in very deed we "walk the hospitals" in order to gain a practical knowledge of disease!

* "In *peripheral paralysis of the facial* the expression of the face is very striking, for, owing to the loss of muscular tension on one side, it falls, whilst the opposite side is drawn up. This distortion is much increased in smiling or talking, or whenever the influence of the will is exerted on the muscles" (Wilks, op. cit., p. 439). But in laughing or talking the impostor moves, he cannot help moving, the paralysed side, and the asymmetry becomes less instead of more obvious. We have a case in mind.

It is singular, when we come to look at it, how ignorant some persons seem to be of the existence of deformities or states of body which must have been endured for long. No better example could be given than that of a shoemaker, 70 years of age, who came to St Mary's Hospital complaining of pain over the right side of the thorax. When he had been stripped it was found that he had lateral curvature of the spine to an extreme degree. He said that this condition was quite of recent date, and that he had only noticed it since he felt the pain. Judged by itself alone his story was enough to cause alarm, but it was clear, both from the ossification of his costal cartilages, and from the fact that not a single cubit had been taken from his stature, that this spinal curvature was old, in all probability nearly as old as the man himself. Had it been an acute change at this time of life, and had the pain which brought him to the hospital been due to the curvature, or been a symptom of some serious malady which lay behind it, a simple liniment and a tonic would not have restored him in a short time to his usual state of health. True, in this case there was no desire to impose; but how readily, in other circumstances, might such a deformity have been used as the foundation of a fraud!

We should not refer to the absence of perfect symmetry between the two halves of the body had we not so often seen some trifling difference in the facial lines, or some excess in size of one limb over the other, and of one half of the spinal muscles over the other half regarded, when taken by themselves, as evidence of very grave disease.

The following case is an admirable example of want of accurate symmetry:—W. G—, a labourer, aged 42, came to the out-patient room of St Mary's Hospital on June 12, 1878, saying that on waking yesterday afternoon from his usual twenty minutes' nap after dinner he found his right arm and hand "numb," his fingers and thumb flexed, and his wrist dropped. We need not dwell on this paralysis of the extensor muscles further than to say that it was only partial, and that it was accompanied by some tenderness above the elbow on the inner side of the arm, and by a small anæsthetic area over the ball of the thumb. As is usual under such conditions recovery was slow, and six weeks after he was first seen, although he was then able to work, he had not regained full use of his wrist and hand. Interesting though this part of the case may be, the special point to which we would refer is the fact that the left side of the man's face was almost entirely wanting in facial lines, while their presence on the right side gave him the appearance of having left facial paralysis. To the palsy of the arm there thus appeared to be added paralysis of one side of the face, and the combination might have suggested grave disease, had not a careful inquiry into every circumstance of the case led to the diagnosis that this asymmetry was peculiar to the individual. And this opinion was subsequently confirmed by the patient himself.

But how easily, under the besetting temptations of railway injury, might a hastily expressed conclusion as to the pathological origin of such a state have given an unscrupulous patient the opportunities of using his natural peculiarity for purposes of deception and fraud. We meet with such conditions every day, and it is of supreme importance that we should duly recognise and rightly estimate them, not only that we may allay the anxiety of the patient who honestly believes his old complaint is new, but also that we may nip the means of deception in their very bud, and not ourselves unwittingly become the impostor's friend.

We must, however, be careful above all things that we do not unjustly attribute to any one the design of deliberate and wilful deception, for, bordering on the very confines of assumed disease, we meet with many of those examples of functional, hysterical, emotional, or neuromimetic disorders, wherein the sufferers are not more likely to deceive others than to deceive themselves. That the manifestations of hysteria are not of necessity feigned or due to expectant attention is shown by the fact that they may be found even in young children who have what may be called the requisite hysterical neurosis. Some remarkable cases of "Hysterical analgesia in children" have been recently recorded by Dr Barlow,* which show conclusively that a symptom of hysteria, which in the adult may lead to the suspicion of imposture, may be seen in children so young that the very idea of feigning is out of the question. The one common characteristic of all these cases, as has been already pointed out, lies in a strange perversion and abeyance of volitional power or will, whereby each action, word, and thought, seem to run riot, as it were, for want of due control. Largely unreal and independent of structural change, the symptoms admit of easy exaggeration and representation whenever the uncontrolled whim of the moment shall allow; but having grown, step by step, out of slighter conditions which there was neither the wish nor the determination to subdue, they seem in their very nature to exclude deliberate imposture. Practically, too, we know that the kind of treatment these functional disorders require is very different from that which suffices to cure ailments altogether feigned. Be the condition, however, what it may, we must remember that to the patients themselves it is very real: the pain, the stiffness, the palsy, are to them as great as they are described, or as full of evil consequences as it is imagined and believed; and though we may regard these symptoms as of little moment in themselves, we must not look upon them as altogether feigned. Here is a good example. A strong and healthy looking girl, aged 16, was admitted into St Mary's Hospital on September 13th. She gave an obscure history of injury to the left knee, but more precise and definite was her story

* 'British Medical Journal,' vol. ii, 1881, p. 892.

that for two years she had suffered from pain in the left hip, and that quite recently she had been six weeks in bed in a country hospital, with a long splint applied for disease of the left hip-joint. While in bed her heel had become drawn up and her foot extended, and she was now hardly able to walk, her left leg being always advanced in front of her, the foot fully extended, and the toes only touching the ground. Examination revealed an entire absence of physical signs, a thing in itself remarkable in hip-joint disease of two years' duration. There was great hypersensitiveness all about the pelvis, thigh, and leg of the affected side, and to a hardly less degree on the corresponding parts of the sound limb. She said she was quite unable to flex her foot, but at the same time she offered active resistance to this movement being made for her. Under chloroform—which, by the way, the malingerer will never take—on September 17th her foot at once assumed a natural position, and it was thus fixed firmly on a splint. Two days, however, had not gone by before her leg had been worked out of the splint and her foot was again in the same position as before. She now begged to be allowed to get up; occupation was found for her in the ward and encouragement was given her that she would get well. Manipulation of the ankle was resorted to, and was usually accompanied by convulsive crying. It was noticeable in these manipulations that the foot could be brought to a state of flexion only when her crying distracted her attention, and that when she left off crying she used every effort to extend it as before. She believed, however, that the manipulation did her good. She improved day by day, began to walk slowly about the ward, went into the garden, and when she left the hospital, on September 30, her gait was natural and free from lameness or suffering. Cases such as this are not uncommon in both sexes after the injury and distressing shock of railway accidents. They may degenerate into wilful deception, for the means and opportunity are ever at hand. But without the most careful scrutiny and inquiry it would be an error to look upon such conditions as generally fictitious or purposely feigned.

Of the cases which have already been given in illustration of the various points and principles which it behoves us to bear in mind, we have designedly drawn the greater number from ordinary hospital work; and they are of value in showing that no special practice is required to teach us many important lessons which bear upon the subject, and that these can be learned from cases which are around and beside us every day.

Turning, in the next place, to the more special topic of malingering after railway accidents, we have to acknowledge, as we had to acknowledge before in speaking of the after-history of nervous shock, that ordinary hospital patients provide little opportunity for gaining familiarity with the kind of malingering with which we have now to deal.* We find, however, that that one element is now obvious, which is so often wanting or obscure in many of the fictitious symptoms with which we may have to do. We refer to the motive which is the stimulus for, and the prime agent in, the assumption of illness and disease. The motive lies in the fact that the law of the land entitles a man to compensation for damage to his person and loss in his business consequent on the negligence of the public company which had engaged to carry him. Has his injury been great; has he lost a limb, or been otherwise maimed; or has his life been then, or in the future, imperilled; it is almost needless to remark that, while no money can adequately compensate him, the amount to be paid him must of necessity be large. When, however, the injury is trivial and passing, when there is neither structural

* They did not, but the Employer's Liability Act is providing the opportunity and temptation, as we have seen in several cases in hospital practice during the last three years.

damage nor prospect of lasting enfeeblement of body or mind, it is evident that the amount of money sufficient to compensate him ought to be very small. And herein, in the endeavour to gain large compensation for small injuries, the malingerer finds reason and excuse for practising deception in order to magnify his claim. The motive is one requiring great moral courage to resist. Many a man whose character has known hitherto no stain, has yielded to the temptation, and has thereby lowered himself in general esteem. Too often have railway accidents had a baneful influence on the lives of men, whose greatest calamity it has been, not that they have received physical injuries which have maimed their bodies, but that their moral natures have been seared and tainted by abandonment to this temptation, the escape from which they would not see. It is not the language of sentiment but of sober fact, when we say that whole households are sometimes made miserable by the devices to which it is needful to resort in order to obtain the desired end. Hopes conceived of future gain, thoughts centred on the one aim in view, and but a sorry consolation when the day of reckoning has come. So powerful indeed is the motive that we find persons taking for long to their beds, abstaining from food, shutting themselves up, neglecting their business, and making themselves weak in body and wretched in mind.

It is very curious how like one another are the means adopted in different cases. Few things are more extraordinary in the history of imposture after railway accidents than the knowledge possessed by persons in the humbler walks of life of the kinds of injury which are popularly deemed inevitable in a collision. Provincial journals often publish in considerable detail the cases and symptoms of those who have, either by litigation or otherwise, received compensation for injuries, and therein may be gained, perhaps, both method and suggestion for adopting a similar course to that which has been successful before. We saw a patient not long ago who was making a large claim for trifling injuries. He was a small publican well up in years, and he had some of the usual symptoms of enlarged prostate. When this was pointed out to him by those who saw him as the real cause of his trouble, he replied, " No, it is my spinal cord to which all my symptoms are due." Are these things learned in books, or whence does the knowledge come? " Drag your leg, you fool, don't you see the doctor coming!" called out by a workman to a fellow who had been in a railway accident, and whom a doctor was on the way across the yard to see, was but the audible expression of many a like lesson which may with greater ease and subtlety be learned or told in the tranquil solitude of a sick room.

And yet it is unusual, we think, to meet with cases where it would be possible to say with certainty that no injury has been received at all. The pain of some trifling bruise or strain is exaggerated and unduly prolonged, and thence are developed other conditions and complaints in whose very obscurity lie the ready means of untruthfulness and deception. The wide-spread, yet erroneous, impression both throughout the profession and the laity that the effects of a railway collision are most likely to be remote, does much to foster a sense of uncertainty and alarm, and to give the malingerer scope for the course which he intends to pursue. Hence it happens that it is after the most trivial accidents, or in cases where no definite injury has been sustained upon which to base a claim, we hear most often of the obscure, subjective, and intangible symptoms and complaints which are supposed to indicate some serious damage to the nervous system, and to forbid all prospect of future recovery. And we ask if this is not a very strange anomaly, something altogether extraordinary, that it is only the slightest injuries which are followed by these purely subjective symptoms, whose very obscurity gives rise to alarm? Is not some light, however, thrown upon them by the facts of accumulated experience, that these symptoms vanish and the complaints are no more heard when the motive for their existence is at an end? Have we not strong grounds for doubting their genuineness and reality? Are we right in ignoring the absence of early symptoms and signs of injury or shock, and in assuming that a condition is alarming, or a prognosis grave, simply from assurances of the patients themselves? And yet this is what we see far too often in dealing with cases of alleged injury after railway accidents. Little regard is paid to the early condition and to the actual state of the patient, and undue attention is given to ceaseless complaints of hidden symptoms, whose real existence we should be all the more cautious in acknowledging, being as they are without objective signs, and traceable to no injury met with in the beginning. Time runs on, the complaints become louder and more continuous; and forensic eloquence, it may be, is left to tell a harrowing tale of the frightful collision, and of a nervous system shattered and beyond repair.

We trust that we are not ourselves lapsing into the "region of *nisi prius*." These are facts, however, and we would that they were more widely known. That cases of severe and unquestionable injury should usually go on to recovery, or to such restoration as the nature of the injury will allow; and on the other hand, that cases of slight injury, hastening in ordinary circumstances to recovery without complication or sequelæ, should after railway collision be followed by innumerable and protracted subjective symptoms on which not a finger can be laid, are facts to arrest attention and to call for explanation. And the simplest explanation is the best. In the one set of cases there is an obvious basis of compensation for the definite injuries received; while in the other there is little or no such basis, and there arise exaggeration, and unreality, and subjective symptoms to make the specious foundation of a claim. But enough perhaps, directly and indirectly, as to the motive for malingering. It is notorious, and it has an all-powerful influence over the course and symptoms of railway injuries.

Let us now learn from definite examples something of the ways of malingerers. A travelling agent, aged 58, received, in a trifling collision, a blow over the right iliac crest. He complained at the time, so he said, to a fellow-passenger of being hurt, and of feeling rather faint. There was never any mark of bruise, and his medical attendant—a hospital surgeon well able to judge—thought his injury altogether trivial. The patient, however, abstained from his work; nor did he resume it until eleven months had passed away. During this time he complained of pain about the right hip which compelled him to use a stick and made him walk lame; of pain in the head; of inability to sleep; of poor appetite and nausea; of constipation; of such general weakness as to prevent him from walking a mile, and that only now and then; of impaired vision so that he could only read the largest type; of loss of memory; and of incapacity to apply his mind to anything, so that neither physically nor mentally did he feel himself fit for any occupation. He frequently stayed in bed for the whole day, and rarely got up before twelve; for days together he never went out of doors, and he took the very smallest quantity of food. Thus he gradually acquired a worn and anxious aspect, and looked pale, thin, and ill. No known means of examination were able to discover any sign whatever of injury or disease; and although he twice undertook journeys of some two hundred miles to be seen by a well-known oculist, could we ever with the ophthalmoscope find a trace of disease in his eyes, or of any cause for his loss of vision. No remedies adopted seemed to have the slightest effect upon him. Liniments did not soothe his pain; soporifics did not make him sleep, and tonics improved neither his appetite nor his strength.

On only one occasion did he give a clue to the absence of perfect genuineness and reality in this case. To a surgeon who had several times examined him by request of the railway company, the patient one day remarked, as he left the house, "You've got a cob-web on your hat, perhaps you'd like to brush it off." He was known to be at this time, and he had been formerly, in pecuniary difficulties. He made a very large claim. This was ultimately settled, and he forthwith went into the country for change of air. He returned home in a fortnight looking, and saying that he was in every respect, perfectly well. He resumed his work at once and has continued it up to the present time, now many years since the accident. Could recovery in this case have taken place in so short a time had the symptoms and complaints not been purposely induced, maintained, or fabricated? Is it not as well-nigh certain as can be that had not the prospect of compensation held out the temptation to this man to make the very worst of his injury, with the hope of pecuniary profit to himself, he would have been laid up for not more than two or three days? And yet he was an invalid for eleven months; a wretched picture, indeed, of induced *malaise*, but a malingerer nevertheless, purposely maintaining his condition in order to increase his claim.

We will now compare two cases out of the same accident, those of a father and son. The father, a man in a small way of business, aged 58, was seen in bed at an hotel in a provincial town the morning after a bad collision. He was unable to give any account of the accident, having been stunned by a blow which had closed both his eyes. He also had a broken rib, and was much bruised elsewhere. He had had about two hours' sleep in the night, had rallied, and felt better. A week later he was still mending, and although he was pulled down by rather severe epistaxis, he was able in twenty-four days to be moved home. There he continued to go on well. In six months his claim was arranged without trouble for a reasonable sum, and when we saw him again three months afterwards he looked and said he was perfectly well.

The son, a strong and powerful man, aged 24, a messenger by occupation, was not stunned, and was able to help his father to the hotel. When seen on the following day he said he had had a good night, but that he felt rather shaken. He had a slight bruise on the right knee, and a simple fracture of the third metacarpal bone of the left hand. A week later, to our astonishment, we found him still in bed, but neither in aspect, temperature, nor pulse was there anything abnormal, and he had been able to eat and enjoy from the very morning after the accident three meat meals a day. He was ordered up, and he would have been sent home immediately had not the illness of his father obliged him to stay. When he got home he at once, and by the aid of a litigious lawyer, took steps to make an exorbitant claim against the railway company. Two months after his return we learned that he had not yet resumed his work. The reasons assigned for this were that he had giddiness and pain in the head, that his memory was bad, and his sleep disturbed by dreams of the accident, that he was very weak and incapable of work, and that he could not trust himself to apply himself to anything. His knee and hand were still in bandages, the arm being carried in a sling. We failed to discover any sign whatever of ill-health about him. Thus he continued to complain and to live in idleness until ten months had flown, when a jury awarded him a sum in reasonable compensation, or about one fourth of that which he had demanded. He was then to every appearance and test in perfect health, and seldom we imagine has so strong and robust a man sworn to so wretched a state of body and mind.

The next is a good illustration of the weeping impostor. A man, aged 42, strong and healthy, was in two slight collisions. In the first there was no evidence or history of his having received any blow or injury. He stated, however, that the day after the accident he felt weak, ached all over, and had pains in the loins and legs as if he had a cold. No amendment took place, nor had he indeed been free from suffering up to the time when we saw him eight months afterwards. He then complained of weakness and fatigue, of bad sleep and loss of memory, and of such great nervousness that the whistle of an engine or the sound of a train threw him into a state of excitement. His manner was whining, and he made several attempts to cry. His medical attendant had never found any evidence of illness beyond the statements made to him by the man himself. Neither in his general aspect, nor in the action of any one of his organs, could any sign of ill-health be discovered. He abstained from work for fifteen months, and at once resumed it on the settlement of his claim, then grown in size through length of alleged suffering and loss. Two years went by, and he was fortunate enough to be in a second collision in which he received a trifling bruise on one leg. He at once abstained from work, placed himself under medical care, made precisely the same complaints as before, and presented as little evidence of ill-health. Again there were months of idleness, and again the settlement of claim and immediate recovery and return to work.

Further examples of this particular kind of malingering are, we think, hardly necessary, although we might give many which, with insignificant variations, have followed the same course. They are often styled cases of "shock to the nervous system." Their chief characteristic is the obscurity, the intangibility, and entire subjectivity of all the symptoms and complaints which disappear at a particular moment. They are not rarely accompanied by some manifest disturbance of health which is induced, we feel sure, by the mode of life to which the individual restricts himself. Can want of proper occupation and exercise act otherwise than harmfully even to a healthy man? Sleep becomes unsound, the bowels are sluggish, appetite fails, the glow of healthy energy and vitality is lost, and there is no longer the picture of perfect health. Be the bodily derangement thereby great or be it infinitesimally small, health and vigour are restored, and work is resumed, as soon as it is no longer incumbent on the man to appear ill and to remain idle, and when the requisite effort to return to a natural mode of life is once more allowed. The man has nursed himself into a state of illness and has thereby generated a condition of "spurious nervous shock," and if he only has the prudence to complain of his back, his case becomes a grave one of "concussion of the spine."

Examples of this induced illness are far from uncommon, and the very fact that the patient seems really ill places an additional difficulty in the way of an accurate diagnosis of the cause. Apart from the necessity of learning what was the precise nature of the accident, and how the man was hurt, it is very essential to know how he passes his time and what are his habits and occupations day by day. It is of paramount importance to separate objective signs from subjective symptoms. The absence of signs and the presence of subjective symptoms alone may fairly warrant a suspicion which ought to entail the most careful inquiry into all the circumstances of the case. Why is that man in bed? why does he stay for weeks indoors? why is he taking hardly any food? why is he pale and thin? why does he sweat?—questions such as these the surgeon should ask himself, and he should not rest satisfied with a tacit belief in what he hears or sees.

We meet with yet other cases presenting combinations of obscure subjective symptoms and of precise objective signs which throw mutual light upon each other, or which perchance may tend to make a diagnosis still more difficult. A traveller by occupation, aged 53, out of employ, took to his bed and called in a doctor three days after a most trifling bump against the stop-blocks in a station. He made the usual complaints of pain, of shock to his nervous system, dimness of vision, loss of memory, and the like, but there was never any sign whatever of ill-health or functional derangement beyond what might be fairly attributed to some atheroma of the aorta. The extravagance of his language and his exaggerated estimate of all his complaints were in themselves enough to raise suspicion as to their reality. Objectively he complained of loss of power in the right arm with stiffness of the right elbow, and of a putty-like sensation of the left leg below the knee and stiffness of that joint. In all ordinary movements, such as undressing or helping himself out of bed, he used the right arm quite as much as the left. It presented no difference from its fellow, though when speaking to him about it he always held it against his side. He resisted with great force when you attempted to bend it, and he called out loudly, as if from pain, when you touched the arm, however lightly, about the elbow-joint. His left leg showed no physical signs of injury or disease about the knee. Held usually rigid and stiff, he resisted any attempts at passive flexion. Flexion to the slightest degree, he said, caused him great agony, although his face showed no sign of suffering. On another occasion he complained of excessive hyperæsthesia of the left knee, however gently you touched it with the finger, although he pulled up and put down his trousers over it with perfect composure. He complained of great agony in his bladder, although he only passed water at natural intervals and in proper quantity, and he could hold it for six, eight, or even twelve hours. He stayed in bed about ten weeks, and took, as he admitted, hardly any food. Towards the close of the twelve months, during which this state of things went on, he had become somewhat weak and thin. He quite early had made an exorbitant claim, and as this was naturally resisted, litigation was in the end the means of decision. While waiting for his action to come on the man was to be seen walking about near the court with the lame stiff leg, and the flexed arm held rigid to his side. His action over, he there and then resumed a natural gait, all trace of lameness having passed away. Were the subjective symptoms less unreal than those which so quickly vanished?

Now let us give another case which offers, perhaps, greater difficulties in diagnosis. Omitting details of the usual subjective ailments, the objective symptoms were frequent vomiting, and such great infirmity of limb that walking was only slow and laboured. The surgeon who saw the man on behalf of the railway company and who gave the history of this case felt sure from all he saw that the symptoms were far from genuine, and among them that the vomiting itself was a deliberate volitional act. It was, to say the least, a suspicious feature in the case that the patient had been seen, when he thought he was out of sight, to start off at a natural pace, swinging the stick on which but a moment before he had been compelled to lean. Knowing all this, the surgeon felt it his duty to tell the private medical attendant what he thought about the case, and he resolved to do so when on the next occasion they were to meet at the doctor's house to see the patient. Having told him his opinion, the doctor pointed out the utter impossibility, in his belief, of so grave a symptom being voluntary, and took him into his own yard to show how the man had vomited since he came to his house not long before. The surgeon's opinion, nevertheless, remained unchanged, and the result of the case justified and confirmed it. Infirm of body and mind, incapable of work, vomiting up to the day of his action for damages, the man immediately recovered when litigation was over. In the case of a horse-dealer who rapidly wasted and became extremely ill, and in whom the most prominent symptom was sweating, the man subsequently acknowledged, when his speedy recovery after compensation excited his doctor's surprise, that he had deliberately sweated himself by violent exercise in thick clothing in order to reduce his weight and size. To those who have never seen such cases it may appear almost incredible that symptoms like these can be volitional and unreal.* Take, however, into consideration every circumstance and feature of the case; learn what has been the original injury; recognise how singular it is that a symptom, alarming in itself, should be by itself independent, and without ill result; remember how powerful is the motive for deception; inquire what steps the patient is taking to gain the desired end, and we have the means, if we will only use them, of arriving at a right diagnosis.

The character of the original accident and injury is far too often forgotten in the later examination of these cases, and a trifling bump is magnified into a serious collision. In a case recently under our observation a man had received an altogether trifling blow on the side, from the manner in which he happened to be sitting, when the train attached a carriage at a station, and the so-called accident ultimately became a severe collision in which the train had been backed into a carriage at the great speed acquired in a run of half a mile, with a crash like thunder. This was the story upon which those were asked to form an opinion who were called in to see him when after several months he had nursed himself into a condition of much weakness, nervousness and malaise, and when his very obvious illness seemed almost to demand a serious accident as its cause. In only one of two ways, it seems to us, can support be given to the presumable reality of the symptoms in such cases: either on the part of the patient by untruth in his account of the accident, or by a blind belief on our part in the

* Some remarkable instances of factitious vomiting are recorded by Gavin, op. cit., p. 256, *et seq.*

all-comprehensive, far-reaching, and inevitable effects of vibratory jar.

We cannot omit a reference to the frequent assumption of injury to the "spine."

We saw not long ago a highly respectable frequenter of the turf who had taken to his bed after a very trifling collision. He was unable to give any account of the accident, *having been knocked insensible and been carried in an unconscious state to the waiting room of a station*, distant a quarter of a mile. He had, however, been able to make a long journey home two days after the accident, that is to say, when the races were over; and he at once took to his bed, and called in a venerable member of the profession to attend him. We saw him on the eighth day. When asked what he complained of, he answered with ready assurance, "Shock to the nervous system, and injury to my spine." He could give no other account than this of his complaints, except that he was wholly unable to get out of bed. Examination, which failed to discover the slightest trace of injury or constitutional disturbance, accidently revealed that this gentleman had a chancre; and this discovery afforded the excuse for promptly ordering him to rise from his bed and walk about the room. His doctor had regarded it as a case of very serious spinal injury. The claim was forthwith settled for a small sum, upon presentation of our report to the company. This same man subsequently appeared as a witness on behalf of a friend, who had brought an action for damages for a like alleged injury in the same accident. He was a very valuable witness, for he swore that he and another man had carried their friend, then perfectly unconscious, from the scene of the accident to the waiting room at the station. The friend, an even more outrageous impostor, had had a slight bruise on one hip. He also finished his mission at the races, came home, and at once took to his bed complaining of his spine. "The pain in his back," we wrote in our report some weeks after the accident, "is so bad that a longer stay than half an hour out of bed is, he says, almost more than he can bear. His pulse is perfectly tranquil, his temperature normal, his aspect not that of ill-health or urgent suffering, and the appetite and bodily functions are as good as can be expected in any one who has so long been without appetite and in-doors. It is our conviction that he is grossly exaggerating his symptoms and complaints. There is no evidence whatever of injury to his nerve-centres—either cerebral or spinal; nor is there any ground for believing that there is any real tenderness of his spine, for wherever we touched him, whether on the spine itself, on the muscles near it, or on the ribs far away from it, there was the same unreal hypersensitiveness and manifestation of suffering. That he has some pain here and there is quite probable; but there is no disease such as to call for his stay in bed or in the house for another day." It need hardly be said that litigation ended this case, which was a highly profitable one for the lawyers.

We have recently (Nov., 1884) learned from this man himself—perhaps in the circumstances the least trustworthy source of information—that his case from beginning to end was a fraud, that he was never hurt at all, and that no amount of money would induce him to go through such a course of illness and confinement again, or to endure such suffering as he had at the hands of those who deemed it necessary to run pins into his legs and to apply the "electric test" to measure his assumed insensibility. We venture to say that no such "tests" were ever needful, and that their tendency was only to confuse and obscure what ought to have been sufficiently obvious to everyone who saw the patient. How fallacious they are, and in this case were, it is superfluous to point out, and yet they were solemnly given in evidence in court as conclusive signs of the serious nature of the man's disease.

Then, again, we have a class of cases wherein the patient

may simulate the mimicries of disease. There are no cases in which it is more imperative to know something of the patient's history, of his previous health, his position in life, his condition immediately or soon after the accident, the nature and extent of the accident itself, and the whole aspect of the case from beginning to end. T. J—, aged 43, was in a slight accident in which he had a small bruise on one cheek and also at the back of the head. He was seen shortly after the accident, but there was no sign either of injury, beyond the bruises named, or of constitutional disturbance. In about ten days he was "taken worse," but in no very definite way. He gave notice of a claim, and then began to complain indefinitely of pain in the back, of pain in the legs, and oppression in the head. There was still, however, no evidence of illness or constitutional disturbance. He continued to get worse and two months after the accident took to bed. He had not been in bed many days when he had a "bilious attack" with constipation and vomiting. This pulled him down and he made no further attempt to get up. A month elapsed, and he then was seized, so he said, with a convulsive attack in which his legs were drawn up, and he was very violent. From that time forward he professed to be troubled with "contractions of the limbs and severe pain in the legs, aggravated by attempting to sit up." He also complained of queer sensations all over, numbness in his tongue, for example, creepings in his legs, tenderness of the palms of the hands. Pulse and temperature alike remained perfectly normal. Five months after the accident he was still in bed complaining of great pain in his back, of pain and tenderness in the legs, and of inability to stand if he got out of bed. He held his hands out somewhat in the position of tetany, but the contraction immediately disappeared when he ceased to direct attention to it. Although when he first got out of bed he allowed his legs to slip away and himself to fall, he only had to be engaged in conversation to show that his legs were amply strong enough to support his whole body. There were no objective signs whatever of paralysis, nor was any illness to be discovered about him except such as might fairly be accounted for by his having been in bed for three months. Thus his muscles were somewhat flabby, his face was pale, and his tongue was furred. Temperature and pulse, however, were normal; his pupils were of healthy size; his mind was perfectly clear. So his state continued until the close of a year, when his claim was settled by compromise on the verge of litigation. We say " on the verge of litigation," for it was held that this man had received a very severe injury to his nervous system, that prospect of recovery was very small, and that it was wholly impossible for him to appear and give evidence as plaintiff at the action. His evidence was indeed taken by commission, the man being quite unfit to leave his bed. He had made a very large claim—not the first in his life, for he had received compensation for alleged injuries some years before. We have heard the history; the absence of serious injury at first; the progressive development of graver symptoms and complaints as time went on; the claim; the examination by commission; the settlement on no meagre basis—for had he not been ill a year?—was he not too ill now to give evidence in court?—what prospect was there of his ever again being well? A very early prospect, as *from the whole history of the case* we had thought it our duty to advise the company. Mark the sequel. In ten days this man was out of doors, in a fortnight he went away for change of air, and in two months he resumed his usual work. He has continued at work since that time in the enjoyment of good health.

In instances such as this it is often difficult, and well-nigh impossible, to say whether the condition is one dependent on a real functional disturbance or is altogether feigned. And the difficulty is largely due to the fact that

18

a functional disorder may be more or less under the control of the patient himself, as we have pointed out on a previous page. It is therefore of greatest importance to search for some symptom which may rightly be placed in the category of hysterical or emotional disorders. Every objective sign that you discover may be under the patient's own control, may be a physical condition altogether assumed, and the clue to the whole case may be entirely wanting until some symptom be found which is outside and beyond his voluntary control or is even unknown to him. Analgesia or anæsthesia for example may form the only indication of the real character of the disease. Dr J. Putnam has admirably dealt with this aspect of the subject in a paper on "Recent investigations into the pathology of so-called concussion of the spine, with cases illustrating the importance of seeking for evidence of typical hysteria in the chronic as well as in the acute stages of the disease" ('Boston Med. and Surg. Journal,' vol. cix, Sept. 6, 1883). After recording two cases of hemianæsthesia in men he says that this is a symptom which shows that the nervous system has in all probability been subjected at some past time to some considerable perturbing influence, and its presence or absence might prove a welcome aid to diagnosis. It may throw light upon the case, but, as we have remarked, it does not necessarily exclude a considerable degree of wilful exaggeration or even downright fraud in the other symptoms and signs. An example has been communicated to us of a deliberate impostor who had lost all tactile sensation over the mucous surface of his nares, a symptom which it would surely be impossible to feign, and which was a clear evidence that some effect had been wrought, somehow and somewhere, upon the nervous system. Regions of anæsthesia, moreover, which follow no recognised anatomical distribution of nerves, say of one leg up to a precise limit at the knee or half way up the thigh, are occasionally present. We may fail to account for them, but they are none the less of deepest interest in a study of the mysterious workings of the nervous system. It is in manifestations such as these, in many wondrous examples of the close interdependence of mental and bodily states, that railway collisions provide subjects of thought for the most learned and philosophical men of our profession; and we do not doubt that much more would hitherto have been made of them if the field had not been too hastily occupied with the meningo-myelitis theory, and had not litigation deterred many from the study of them. The meningo-myelitis theory has done serious harm in more ways than one.

We have nothing to do here with the pathology of morals, nor need we essay to gauge the different degrees of moral obliquity in undoubted feigning and assumption of disease, and in wilful exaggeration of real conditions as a means to compass some aim in view. Were we to include all cases of simple exaggeration under the same head as cases of fictitious and feigned disease, the material would from very bulk become unmanageable. If, as we have said, the motive be so strong and so prevailing, it is natural and only human that we should meet with exaggeration in a very large proportion of the persons injured in railway accidents. But, on the other hand, it behoves us to remember that exaggeration may not be, nay very often is not, altogether wilful or assumed. Exaggeration is the very essence of many of those emotional or hysterical disorders which are so common in both sexes after the shock of collisions. Here it may be an idiosyncrasy of the individual; there it may be the outcome of mental disturbance from the fright and alarm amid which the injury was received. It is only by a consideration of every feature and aspect of the case—clinical, pathological, social, and moral—that we shall rightly estimate the kind of exaggeration or malingering with which we have to do.

Do we yet need further help to lead us to a right answer to the question so often forced upon us: the whole bearing of the individual patient must be carefully observed. It is a fact which calls for no special skill to recognise that the malingerer reveals himself as oft in word and manner as in deed. His very speech bewrays him. He is extravagant in language and utterance; his every complaint is told in words which are needed to convey significance of evil; he displays a bitter animus against those who have injured him, who have rendered him unfit in the future for work or happiness, and who have doomed him to a life of wretchedness and poverty and disease. There is no gleam of hope in his condition, no prospect of enjoyment of life again. All this from day to day, from month to month, without change or variation, until the acme of his settled claim.

> "Before the curing of a strong disease,
> Even in the instant of repair and health,
> The fit is strongest: evils that take leave
> On their departure most of all show evil."
> (*K. John*, Act iii, Scene iv, 112.)

Or yet again we meet another type, more odious, but not less easily perceived.

> "There is no vice so simple but assumes
> Some mark of virtue on his outward parts."
> (*Merchant of Venice*, Act iii, Scene ii, 81.)

With glib facility of tongue he talks of the frauds which are so notorious upon railway companies, but his own character is, and always has been, above suspicion. His complaints are many and grievous, but yet he would not make them worse than they are, bad enough though they be to keep him from his work which his doctors urge him to resume. Occupation is impossible; he cannot even leave the house; and his religious sense is shocked that for so long he has not been to church. He can bear no noise. He cannot read, and his only diversion is to hear his Bible read to him by his children or his wife. He is pleased to see you, for he knows how deep and true an interest you take in his wretched state; and he is ever ready to fall in with—but not adopt—the suggestions you may make for his comfort and the improvement of his health. Once more, as you leave him, he assures you, with Pharisaic unction, that he is not as other men, and that he would be the last to try and make money out of the affliction with which he has been visited. His speech again bewrays him, and exposes the pious fraud.

*" The philosophical observer who has given close attention to the extremer forms of Protestantism in their relation to character, such as are known as Evangelicalism, must have noticed how often they go along with an extraordinary insincerity or actual duplicity of character. I mean not to insinuate that the tendency of an evangelical faith is to engender duplicity of character; the reason of the connection probably is that persons of that character are attracted naturally to a form of creed which, making large use of the sort of emotion that springs from self-feeling, yields them the gratification of a suitable emotional outlet, and by the habitual employment of a conventional religious phraseology, keeps out of sight, or at any rate veils thickly, the gross variance between high profession and low practice which the use of a common language could not well fail to bring clearly home. They use conventional language without ever sincerely analysing its meaning, because they find in it fit expression for certain narrow feelings that have been associated with it, and are more comforted by the phraseology than if they really understood it; it has become a shibboleth to them, the sign of special grace, like that blessed word Mesopotamia, the sound of which yielded so much comfort to the old

* Maudsley, op. cit., p. 149.

woman of the village. They are not the conscious hypocrites which they seem; they are inconsistent without really feeling their inconsistency; the two diverse developments of their nature do not interwork, and they go on with an incoherence of character which they never realise."

Thomas Hood knew something, we think, of this character, or he could not have written the following verse in his brilliant "Ode to Rae Wilson, Esquire."

> "Behold yon servitor of God and Mammon,
>   Who, binding up his Bible with his Ledger,
>   Blends Gospel texts with trading gammon,
>   A black-leg saint, a spiritual hedger,
>   Who backs his rigid Sabbath, so to speak,
>   Against the wicked remnant of the week,
>   A saving bet against his sinful bias—
>   'Rogue that I am,' he whispers to himself,
>   'I lie—I cheat—do anything for pelf,
>   But who on earth can say I am not pious?'"

CHAPTER VIII

CONCLUDING REMARKS

It remains for us finally to speak of the collateral circumstances which tend to place in a class by themselves the cases we have considered, and which are almost inseparable from all injuries which become the subject of medico-legal investigation. It must, indeed, be very obvious that railway injuries present many important features which are wholly absent from the ordinary accidents and diseases met with in every-day practice. First and foremost is the fact that compensation in money is paid for the injuries and the consequent loss sustained; and secondly, that the injuries and the loss may become the subject of a medico-legal inquiry which may materially modify the clinical aspects of individual cases, and which unquestionably entails at the same time certain duties and precautions upon the medical man.

At first sight, perhaps, it is not very clear why pecuniary compensation should be an element of importance in the course and history of the spinal and other injuries received in railway collisions. Let us picture, if we can, the change which would come over our hospital patients were a pecuniary value to be placed upon every injury they sustained. How great is the probability that many of them would see only the worse side of their ailments, would lay undue stress upon the pains they suffered, and would exaggerate the term of prospective disablement from work, were these to be essential factors in calculating the money-worth of the injury received. Instead of an earnest desire to regain health and strength and to return to work, hesitation and lack of desire towards these ends would be almost surely met with were the patient's sufferings, their present and future disablement, to count in the sum total of the pecuniary value of their injuries. Were elements like these to enter into the history of our hospital patients the features of many a case would be completely changed, and the whole clinical history of disease would wear a very different aspect from that which we usually see. Illness or injury has prostrated a man, and when the time has arrived for him to profit by advancing convalescence, some effort must of necessity be made by him to resume his former life. The habit of work, and the daily routine of his life, have been broken by his illness, and voluntary effort, stronger or weaker as the case may be, is needed for him to regain his former activity of body and mind. As a bar and hindrance to this natural effort —wholly unconscious in many instances—compensation for his injuries exerts an influence on the course of his progress towards recovery. We say *wholly unconscious*, for the knowledge that compensation is a certainty for the injuries received, tends, almost from the first moment of illness, to colour the course and aspect of the case, with each succeeding day to become part and parcel of the injury in the patient's mind, and unwittingly to affect his feelings towards, and his impressions of, the sufferings he must undergo. He is less likely to take a hopeful view of the future, is more prone to believe that he will be long prevented from following his ordinary employment, and very small effort will be made towards resuming his work when convalescence points to the fitness of his doing so. It is well that this element enters in no wise into the history of our hospital cases; our difficulties in the treatment of disease would be manifold indeed. Our hospital cases, however, are fortunately free from this incubus, and we shall do wisely, therefore, in dealing with the cases involving medico-legal inquiry which may fall under our notice, to revert to our hospital experiences, and to recall the usual history of our hospital patients, both as to suffering and bodily disablement, and the prospects of recovery.

Even in perfectly genuine cases—and it is these to which alone we now refer—compensation acts as a potent element in retarding convalescence, as evidenced in numberless instances by the speed with which recovery sets in as soon as the settlement of pecuniary claim has been accomplished. "He got well as soon as his case was settled," is heard so often after the close of medico-legal inquiry, that we might almost be driven to the conclusion that injuries were not in reality, or but very rarely, received in collisions, did not experience abundantly prove that even when there is very serious disturbance of health the removal of this source of mental embarrassment and worry has a very healing influence indeed.

Do not let us be misunderstood, nor let it be thought that we are endeavouring to lay down a doctrine so absurd as that settlement of claim can of itself be a curative agent, in the sense that it can hasten the setting of a fracture, remove the pain which is an inseparable concomitant of sprain of the vertebral muscles and ligaments, or restore the nervous tone which has been upset by the shock of a collision. The natural forces here, as elsewhere, tend to restoration of health; and recovery is, happily, as perfect after the injuries so commonly received in collisions as after any other kind of injury which the surgeon may be called upon to treat. There is this difference, however, that when the immediate effects of injury are passing away, and every sign points out that convalescence has set in, compensation holds out an inducement to the patient not to make the requisite effort to resume his work and ordinary avocations which are in themselves, as we so often see, the best means of crowning the period of convalescence and of restoration to health. Depend upon it, had the ordinary injuries of hospital and every-day practice to contend with this element, convalescence would be very much retarded and prolonged, and with no less certainty may we say that were this element absent from the contemplation of the injuries we have considered, these in their turn would be recovered from with much greater promptitude than they at present are.

It is only human nature that compensation should have this influence in delaying recovery, even amongst those who have been genuinely injured, and who are honestly desirous to get well and to receive no more compensation than is their just due. A man who would scorn the imputation of using his injuries for the purpose of pecuniary gain, or who might equally scorn the suggestion that he would be a good deal better if he would arrange his claim, may yet be—and very frequently is—the unconscious victim of this element which exercises its influence in almost every case to a greater or a lesser degree. As far as symptoms guide us, no boundary line can be drawn between the conscious resolution to make no effort to put forth natural strength until money shall have been paid, and the unconscious yielding to the influence which prospective compensation exerts. The line very often can be shown only by the nature of the demands which the patient is making, or by the manner in which he is pressing and supporting his claim. These are matters with which we, as medical men, have nothing to do. Our business is to look at the clinical features of the case alone; and when we see the early manifestations of injury have passed away, and that, notwithstanding the expressions of the patient, there is yet distinctive evidence of improvement and of returning vigour, and that at the same time there is no obvious desire, by deed rather than by word, to resume the ordinary pursuits of life—that the case is, in fact, not pursuing the course which we should have expected from our knowledge of other injuries of a like kind—then we may feel sure that the matter of compensation is in some degree present in the patient's mind, and that settlement of his claim is the one thing needed to bring about complete and early restoration to health. Doubtless there are exceptions, and we shall meet with cases where recovery is slow without the clogging influence of pecuniary compensation, and also cases where recovery is natural, and in a natural time, even though compensation may largely engage the attention of those who are seeking it both during convalescence and after restoration to perfect health. The borderland of conscious and unconscious yielding to this influence of compensation is ill-defined; but it is no part of our duty to express an opinion as to whether the patient is on one side of the boundary or the other. All we have to do is to recognise in our estimate of the future, and to apply this clinical fact, infallibly established by numerous cases, that when we meet with evidences of delayed recovery, of an arrest in the natural processes of convalescence, and an absence of any real determined effort to throw off the invalid state, then, truly, settlement of the claim is the one thing which the case requires, and which will do more for it than all the remedial measures which art and science can command. If we do not bear this fact in mind we shall go wrong in our estimate of the prospects of recovery, and we shall fail to hold the balance evenly between those who have to give and those who have to receive compensation for the injuries sustained. Let us illustrate what we have been saying by an actual example; we might choose many cases of a like kind.

W. A—, a stout elderly man, was in a rather severe collision. He was able to continue his journey after the accident, but in two or three days he began to have pains in the back and to feel himself shaken, weak, and ill.

He then returned home, a journey of four hours, and went to bed. There were no signs of injury other than those of slight sprain of the muscles of the back, with general nervousness and loss of tone. He steadily improved, and in two months it was thought that he was sufficiently well to arrange and settle his claim. He then, however, began to complain more; and four months after the accident he looked worse than hitherto, appeared ill and anxious, expressed himself unable to leave the house, and wholly unfit even to think of resuming his business. This went on for several months, and, instead of any improvement taking place, he began to look more aged and worn, and not having been out of doors for long, lost appetite and weight, became prone to cry, and altogether presented an aspect so unhappy that an opinion was given that he was permanently injured from "concussion of the spine," and would never be fit to do anything again. There were, however, no symptoms of serious illness or disease, and settlement of his claim was confidently anticipated as the one thing essential to restore him to health. Nine months passed without a shadow of improvement, and his claim was at length arranged. In a very short time he was perfectly well, looked in good health, and "ten years younger" than before his claim had been settled. Nor was this improvement transitory. He continued in perfect health, and five years after the accident was following his occupation with his usual vigour and in his normal state of health.

A case, it may be suggested, devoid of all colour, and wanting any tangible symptom at all, if we exclude the pain in the back which was pretty sure to conjure up "concussion of the spine." But it is just these colourless cases which are often the most difficult of diagnosis, especially to those who have never seen anything like them before. "I cannot make out why this patient does not get well. He seemed to mend at first, and I thought he was going to get over it soon, but now he appears to have no energy at all. My medicines don't do him any good. I urge him to make an effort to get about and try a little business, but he says he cannot ; or if he does that he breaks down again almost before he has begun." In words such as these have we often heard the plaint of the doctor. They would perhaps be heard not quite so often were this clinical fact in the history of these injuries more fully recognised, that settlement of claim is frequently the most important agent to bring about recovery.

We need not seek too closely to inquire into the *rationale* of the change which may be thus induced. There is a release from the mental worry and annoyance inseparable from any long dispute, and a feeling that now at last, when the whole trouble is over, a fresh start is possible, and that the springs of energy being loosened from the bondage that enthralled them, some persistent effort may once more be made to move about and resume work. And there may be—we will not say there is—a satisfaction akin to joy in placing a good round cheque to your balance at the bank, which, in this age of progress and poverty, exerts a stimulus which no pharmacopœial preparation can supply. The attitude of the patient's own thoughts is wholly altered. Before compensation was effected he held out a goal to himself and said, and thought:—" When my claim is settled I will try and resume work ; but I will wait and see how I am." When his claim has been arranged he argues thus:—" Now I must begin my work again, and do the best I can." The effort, at all events, can at length be made, and with each succeeding hour of activity and occupation there grows up a healthier tone, and the rôle of the invalid is laid aside.

It is not, however, in cases like these alone where settlement and repose of mind conduce to speedy recovery. The same result may be seen elsewhere, and it is a well-known fact, communicated to us especially by two surgeons of large experience, that the health of prisoners in gaol is unquestionably the better after they have received sentence, even though the sentence may have been far more severe than had been expected.

While, however, we cannot ignore the influence of this restorative agent, we must be more than ever careful that our knowledge of its efficacy be not improperly applied. An accurate history of the injury and of the accident in which it was received; a careful observation of the patient's condition immediately after the accident, and of his progress in the days and weeks which followed; an impartial comparison of his case with others of a like kind which we have met with elsewhere—these things will guide us to a right estimate of the facts when recovery seems to be unduly slow, and when examination reveals no signs of disease to account for the delay. And although it does not fall within our province to deal in any way with the matter of compensation, it is yet necessary for us sometimes to know what steps the patient is taking to obtain compensation, and how far this question may presumably be engaging his mind.

While, then, compensation has this influence upon the recovery of patients about whose *bona fides* there may be no suspicion, we have to remember that it may act in others as a strong temptation to wrong. Hence arise some of those difficulties which surround the clinical inquirer, and which entail obvious duties and precautions on the medical man. It is our duty to hold the balance as evenly as may be between the two sides which are more especially concerned in every medico-legal inquiry; and it behoves us, above all things, to be careful that no affection of coincident origin, nor any precedent deformity or disease shall, through ignorance or carelessness on our part, be allowed to form an item in the claim which the patient may think fit to make for the injuries he has sustained.

Nor can we tell how readily a patient may adopt a suggestion which has been unwittingly put into his mind, or how soon an unguarded word or opinion may give a tone to the symptoms which of themselves they do not possess. Let us therefore avoid as far as possible all leading questions, for in the use of them we may give rise to symptoms which had no previous existence. In one of the grossest cases of fraud which has fallen under our notice, a grave opinion was formed of the case because a leading question prompted an answer which was wholly untrue. The man complained of giddiness; and as vertigo was known to be often the result of diplopia, the patient was hastily asked if he saw double. "Yes," was the reply, and led on by further leading questions, a very serious train of alleged symptoms was revealed. Not one of them was true.

The well-known and wonderful impostor who, some years ago, went the round of the hospitals, and who with varying degrees of success simulated many nervous disorders, lacked in his marvellous exhibition of tetanus the usual rigidity of the abdominal muscles. A remark was made at the bedside, when he was in one of the worst of his seizures, that it was strange the tetanic spasm, so extreme elsewhere, should not also affect the muscles of the abdomen. The next day they were as hard as a board.

Not less important, also, is it to avoid the unnecessary use of leading methods of examination. We would not in the smallest degree detract from our instruments of precision, but here we speak of their "unnecessary use," because the cases are singularly few and far between which demand the whole *armamenta* of the specialist, who with dynamometer, æsthesiometer, audiometer, lenses and battery, finds out some trifling departure from the normal which may be made the unfair groundwork of a claim.* Of

* " Were a pathologist," writes Dr Moxon (" Croonian Lectures,"
' Lancet,' vol. i, 1881, p. 568), " with a great microscope to spy through

what earthly use are observations by the dynamometer, unless you know that the patient is telling you the truth ? You discover, forsooth, that he squeezes with a force of 50 lbs. with his right hand and of 10 lbs. with his left, and you forthwith create the scientific and incontrovertible basis of a lie. We speak plainly and with due deliberation upon this point, for nothing has struck us as more extraordinary in our experience of railway injuries than that, in the examination of them, all common sense, the best and surest diagnostic guide, should be so often abandoned, and reliance should be rather placed on methods of examination which are of scientific value only when every suspicion of exaggeration or imposture can be put away.

And amongst leading methods of examination there is none more fallacious or untrustworthy in the hands of those who lack experience than that which we so frequently hear called the "electric test."

Electricity, whether as a therapeutic or diagnostic means has not yet achieved that status that it can be regarded as a scientific "test" in practical medicine. We have no desire to depreciate its real utility, but we are within the truth when we say that its remedial value is a good deal less, and its range as a diagnostic means is far more limited than is very often supposed. That *positive* aid in diagnosis may occasionally be given by it is not of course to be denied. The reaction of degeneration, or the different reaction of *various* muscles in the *same* region, may be, for example, important points in diagnosis. Dr Buzzard has recorded cases of the kind. In a passage, which may well serve to indicate some of the difficulties involved in the application of electricity, he writes:—" We *never* see in hysteria the various muscles of one limb showing differing

all our brains as we sit here in states of satisfaction, he would certainly see a great deal in the way of tortuous capillary and dots of yellow pigment—a great deal that under the microscope would look very alarming."

degrees of abnormality in their response to faradaism, from a condition of total absence of reaction in some, to nearly a natural state in others. Moreover, in hysteria, according to my observation, applications of electrical stimulus (and especially of the voltaic current) on one or two occasions usually suffice to restore the natural excitability of the muscles (equally in all) which has simply declined through disuse. A difficulty can only arise where the observer has but one opportunity of testing the electrical condition, and it is then quite possible to occur. It must be remembered that, as a distinct lowering of faradaic excitability almost invariably signifies organic change in a nerve-trunk or centre, a diagnosis of hysteria can never safely be arrived at whilst that condition persists. On the other hand, I need scarcely remind you that the preservation of a completely normal faradaic excitability in the muscles of a limb does not show that that limb is not paralysed. . .

. . In cases of paralysis, it is only when the integrity of the grey matter of the anterior horn is disturbed, or when there is some lesion of the anterior root or trunk of the nerve, that you find decided loss of electric excitability. You frequently meet with complete paraplegia and yet all the muscles will respond normally to electric currents."* It is worthy of note, however, that in these cases of railway injury electricity is regarded as a conclusive "test," because the patient has shown *no response to it whatever—neither movement nor sensation of pain*—even when its application has been, from the standard of common experience, most severe.

There are fallacies in its application which it is essential to avoid. It is a fact that when patients have been long confined to bed, and from one cause or another—be it from genuine fear of moving, from a supposed inability to move, or from a resolve not to move—have kept their

* 'Clinical Lectures on Diseases of the Nervous System,' by T. Buzzard, M.D. London : J. and A. Churchill, 1882. P. 100.

19

legs at rest, there is considerable diminution or loss of cutaneous sensibility. The "electric test" is applied, and it is discovered that there is no muscular response to the current and no sensation of pain. On this *negative* evidence an alarming diagnosis is at once established; for this "electric test" is assumed to be conclusive upon the usual point at issue—the existence or absence of real organic disease. Coupled, however, with this condition of cutaneous insensibility—and it would be a grave one were it not so—the reflexes are normal, there is no trophic disturbance in any part of the limbs, there is little or no impairment of muscular nutrition, and upon careful and proper examination the electric reaction of the muscles is found to be natural.

On careful and proper examination by a competent observer: for while the skin becomes insensitive to ordinary tactile impressions, it is at the same time a very bad conductor of the electric current, which may therefore, in careless examination, never reach the muscles at all. This fact has been well pointed out by Dr Buzzard, who, after describing a case of hysterical paralysis, writes:—" In this case, as I have seen in many belonging to the class of hysteria,* the epidermis, which had arrived at extraordinary thickness, apparently from disuse of the limbs, offered great resistance to the passage of electric currents. In these circumstances a more than usual amount of care in thorough soaking and rubbing of the skin, as well as in selecting the motor point, is necessary in order to avoid fallacies."†

" As well as in selecting the motor point "—we rewrite Dr Buzzard's words. Too often has the " electric test "

* The very cases, that is, that are so often seen after railway collisions, functional disorders of motion and sensation, and occasionally real imposture.

† Op. cit., p. 118, Lecture v, "On the Differential Diagnosis between certain hysterical conditions and Myelitis."

been applied—in all honesty we are perfectly ready to admit—by those who have never heard of "motor points," and who seem to have thought that the right diagnostic method of electricity consisted in the application of the poles to any haphazard points upon the patient's limbs or body. And this is expert scientific evidence in courts of law! Here let it once for all be said that the application of electricity in disease is one of the most difficult things in the whole range of clinical investigation; that there are very few men indeed who have had the requisite experience or opportunities for its employment; and that without these opportunities and experience, which alone can teach the needful care and skill, no man has a right to come forward and proclaim an "electric test" of the existence of disease. There is no "electric test;" none, at any rate, has yet been found, which is not rather a test of the credulity of him who trusts it.

We have heard it said by counsel, sore-pressed in the heat of litigation, that medical men have no right to question the statements made to them by patients, and that their duty is to hear and implicitly believe the stories which are told. But we would point out that such a doctrine strikes at the very foundation of the clinical investigation of disease. Rarely does a case come before us in hospital or in private practice in which it is not necessary to subject the patient to questions as to the character of his complaints, and, if need be, to cross-examine him as to those facts and features in his history which may not be perfectly clear. We say it with all respect, but counsel might, we think, learn much of the art of cross-examination from medical men. Nay, we will go so far as to say that it forms so common a part of every inquiry into the clinical history of injury and disease, that cross-examination is a more perfect method of investigation in the medical than in the legal profession. Cross-examination with us is invariably used, we hope, in a scientific spirit—solely for the eliciting of truth. May it not be said with all fairness, will they not themselves acknowledge, that with our friends of the long robe, the art and skill and method of inquiry and cross-examination consist sometimes in an endeavour to make awkward facts fit together that, in the stress of advocacy, the real truth may be distorted or concealed ? And what has truth to fear from inquiry or cross-examination ? No true story of injury or suffering, no true symptom of disease need dread the cross-examination of a medical man; every fact, each symptom, will only stand out more manifestly true.

> "A rotten case abides no handling."
> (*King Henry IV*, Pt. ii, Act iv, Sc. i, 161.)

Medical men have a right, and they must exercise it, to subject all cases of injury to the most careful inquiry and investigation; and they would do very wrong to abrogate that function of legitimate investigation which lies in questions and cross-examinations. It is the habitual method of their inquiry, and no complaints of counsel should compel them to abandon it in those cases which may unfortunately chance to be involved in litigation.

It will fall to our lot sometimes to give evidence in courts of law when litigation has become necessary to settle the money dispute which has arisen as to the value of the injuries received.

The day has not yet arrived—it may perhaps be well for our profession when it has arrived—when the evidence of medical men shall be recorded in some other manner than before a jury in open court.

In common with many others we read with deep regret the remarks which a learned physician addressed not long ago to an assemblage of students on the question of giving evidence as scientific witnesses in courts of law. Instead of telling them that medical evidence was so hopelessly partial in railway cases that he himself had declined for many years to take any part in such proceedings, Dr Russell Reynolds would have done better, we think, had he told his hearers how they might best avoid the snares and dangers which unquestionably beset the medical witness in cases involving litigation. Doubtless if a man feels that the cast of his mind prevents him from being an impartial observer of facts, or an impartial witness and exponent of their reality and meaning, he will do wisely, both for his own sake and that of his profession, to avoid all cases where litigation is involved; but to avoid them because others err is a weak way to improve our morals or to advance the truth.*

There must have been few who heard Dr Reynolds who will not very surely be called upon at some time or other to appear as medical witnesses. It is so well-known, he argued in legal circles who amongst the physicians and surgeons of the day will give evidence on this side or that, that the phrase "'So-and-so is a very rising witness' has been used and not unfrequently. "What does this mean," he asked, "but that members of our profession, whose one object should be truth and justice, 'take sides'?" We answer that it does not of necessity mean anything of the kind. Differences of opinion there must ever be in matters so difficult as the investigation and interpretation of disease ; but it is wholly possible—there are instances of it every day—to be impartial, and to give perfectly impartial evidence in the witness box.

There is no reason why a medical witness should not

* "Specialism in Medicine," 'Lancet,' vol. ii, 1881, p. 657. "There are some members of our profession," Dr Reynolds says, "who have become specialists in this direction, who seem to think that everything that a man tells them of his subjective symptoms are matters of fact and of great importance ; and, on the other hand, there are those who regard every plaintiff as either a knave or a fool, and most probably a combination of the two, but who never believe that any man is injured in a railway accident unless he has broken his neck, or has a compound fracture of his thigh."

endeavour to do the work required of him as honestly, as carefully, and as well as any other work he may undertake; and good qualities, we may feel sure, in a "rising witness"—patience, impartiality, good temper, a sense of justice, and a single purpose to give utterance to the truth—are not held in light esteem in the Courts of Justice, although elsewhere they may not save him from reproach.

The warfare of contending counsel, it is true, may throw into strong relief the slightest differences or shades of opinion which may be entertained amongst us. We need not, however, be disconcerted by the hackneyed taunt that "doctors differ," if only we have done our duty by our profession and by ourselves. There is no place where the attitude of a medical man towards a case will be so surely revealed as in a court of law, where he has to give public utterance to his opinions, and where his every word and manner are being closely scrutinised and observed. Not one whit less than judge or counsel does he there hold in his hands the honour and dignity of his profession, and it should be his earnest thought that both should remain unsullied while he has them in his charge.

Better than any words of our own, with which we may well conclude, are the following remarks of Chief Justice Clinton upon the evidence of medical witnesses in courts of law:—*"As to the delivery of testimony by you as experts, I have very little to say that might not just as properly be said to a witness who is called to testify only as to the facts of the case. The difference rests in this; the expert, as such, is asked only for his opinion upon the facts. He may be asked his opinion upon a hypothetical state of facts, and required to give reasons for the opinion he expresses. The cross-examiner is allowed great latitude, and I am sorry to say not unfrequently abuses it. But if the witness will only remember the worth and dignity of

* 'Buffalo Medical and Surgical Journal,' Jan. 1, 1880.

his profession, and that he is there simply to speak truth, as a servitor of justice, no arts or sneers of counsel can disturb him. Calm and self-possessed he will answer every question, direct or cross, fully, and in the plainest and most lucid language in which the meaning of the answer can be conveyed to the jury. To such an answer he will add nothing, unless it be a necessary explanation. He will not air his learning before the court, nor have any the least contention with counsel. The court will, if need be, protect him from the abuse of lawyers. Such a witness will return from the stand as calmly as he went upon it, approved by his own conscience, and respected by the court, the jury, and the bar."

APPENDIX

THE following table contains, inclusive of those recorded in the text, 234 cases. They are not selected. It would have been as easy to put together two or three hundred cases of recovery, and omit those where recovery had not been complete, but we thought it better and fairer to take the cases as they came. For this purpose we have chosen the first 250 cases seen, and have excluded therefrom those cases where injury had been sustained in some other way than in collision. There remain 234, and we stop at this number, not because the 235th case tells a different tale, but because 234 are surely sufficient. Precisely the same lessons are taught by the second series of 250 cases, or by the third, which we have seen; although it is right to say that in these later series there has been a much smaller proportion of cases presenting unquestionable lesion of the central nervous system.

In describing the nature of the accident and estimating its severity we have been guided by the official reports, the number of persons injured, the damage to rolling stock, &c., &c. The terms used are arbitrary, but they will convey, we think, a not inaccurate impression of the various casualties. The date of settlement has been recorded as giving in most cases a rough idea of the length of the illness, or at any rate of the time after the accident when the patient had sufficiently recovered to arrange his claim without injustice to himself. It will moreover be seen that in many instances delay in settlement, and consequent protraction of illness, have been due to the largeness of the claim made which led thereby to long dispute.

We have endeavoured to learn something of the after-history of the patients at a period not less, in many cases much longer, than two years after the accidents. The reports of their then conditions have been obtained in the vast majority of cases from medical men, in a few instances from the patients themselves, and some cases we have ourselves been able to examine. The labour of this inquiry has not been small, and had it not been for the kindly help of many professional brethren, the inquiry must have been futile or impossible. We wish the table could have been more complete and better than it is, but the migratory habits of persons in the poorer walks of life have prevented the success of repeated inquiry and search.

No. of case.	Sex.	Age.	Nature of accident.	General outline of case.
1	M.	21	Carriages ran down embankment	Small wound at back of head. Momentarily stunned. "Felt regularly numbed all down the back and legs," and for a few days had "tingling" sensation in one foot. Pain and tenderness at 4th dorsal vertebra. Later had "hysterical fits"
2	M.	54	Sharp collision	Dazed and sick. Bruised about legs. Soon suffered from great nervous agitation, shown by chorea-like movements of trunk, face, and limbs, with indistinct articulation; this passed away, and five months after the accident only occurred when excited. Complained of numbness in the left arm, which was colder than the right. Had suffered previously from sciatica and rheumatism, and now has aortic regurgitation
3	F., single	48	Same as 1	One rib broken, and bruise of thigh and forehead, and general shake. Subsequently very nervous, and had much pain and hyperæsthesia at the parts bruised, and also in the small of the back
4	M.	32	Same	Much cut and bruised about face and head and the calf of one leg. Had concussion of the brain and considerable collapse. Recovered naturally, but at end of six months complained much of hyperæsthesia of the scalp; he appeared, however, to be in good health
5	F., married	35	Very slight collision	Had a blow on the head and a strain between the shoulders and in the lumbo-sacral region. Was confined to bed about a week with headache and sleeplessness, then gradually improved; but four months after accident complained of considerable "nervous" symptoms and dread of paralysis, with sensation of weakness in left arm. No objective signs
6	M.	63	—	See page 283
7	M.	24	Very slightest collision	Showed no signs of injury, and amused himself by snow-balling while the train was being got ready again. Subsequently complained of his back, and began to walk almost doubled up. There was no sign whatever of constitutional disturbance or disease

APPENDIX

Date of settlement.	When last heard of after accident.	Condition when last heard of.	Evidences of injury to spinal cord or membranes.	Remarks.
3 mos.	6½ years	Was at work again five months after accident, though suffering from pain in the back when lifting heavy weights. Good health ever since. Married and has a family	See outline	A delicate "nervous" youth. He said the accident had "played on his nerves."
6 mos.	14 mos.	Soon after settlement began to do some work, and was slowly improving as far as the nervous symptoms were concerned. Was found dead in bed	None	No autopsy. Bad subject to meet with an accident. Cardiac disease at time. A question whether the chorea and the weakness in one arm may not have been due to cerebral embolism.
6 mos.	—	No record obtainable	None	At the menopause. Had a large uterine fibroid, to which the lumbar pain was in great measure due.
6 mos.	7 years	In good health	None	Made an exorbitant claim.
10 mos.	—	Cannot be traced	None	A nervous hysterical woman, separated from her husband. Shortly afterwards left the neighbourhood. Made an exorbitant claim.
—	—	—	None	
6 mos.	—	Cannot be traced	None	An exorbitant claim. Slight injury and subsequent imposture

APPENDIX

Age.	Nature of accident.	General outline of case.
37	Rather sharp collision	Had a blow over the right zygoma, followed by difficulty and pain in mastication. Headache and sleeplessness. Right membrana tympani supposed to have been ruptured. Ultimately healed
32	Same	Had a blow on the head and a sprain of the lumbo-dorsal region. Suffered much from menorrhagia. Had previously had prolapsus uteri which was increased by the accident
70	Slight collision	Had a blow on one elbow and in right loin; unconscious from fright for ten minutes. Soon began to mend, and recovered rapidly
32	Same as 8	Had a severe blow over the right sacro-iliac synchondrosis. Suffered subsequently from much pain and stiffness at the part. Also from a "general nervous shock," inability to apply himself, loss of memory, &c.
25	Same as 1	Had a small scalp wound which healed at once, and a blow on the knee which threatened suppuration. Subsequently hysterical. A frail and delicate girl. Her knee kept her on the couch for some weeks
62	Severe collision at night	Thrown down. Dazed. Soon able to help others. In three days had pains about neck and left shoulder, and felt generally shaken. Also felt his left arm "cold"—subjectively only. Very nervous, and suffered a good deal of vertebral pain
36	Same	Not conscious of injury, but "dazed" from fright. Said to have been sick a few hours afterwards. Later, complained of pain in lower part of back, with tenderness in the lumbar and cervical spinous processes. Said he had a sensation of "numbness all over." Never any constitutional disturbance
26	Same	Blow on face and back. Helped others. Vomited two hours after accident. Subsequent headache and sleeplessness, and "general nervous shock." The description of later symptoms elicited by leading questions. Called "concussion of the spine." No after-signs of constitutional disturbance

APPENDIX 301

Date of settle-ment.	When last heard of after accident.	Condition when last heard of.	Evidences of injury to spinal cord or membranes.	Remarks.
4 mos.	5½ years	In perfect health and at usual occupation	None	
8 mos.	5 years	In perfect health except in so far as uterine displacement gives trouble	None	
6 weeks	5 years	No subsequent illness, and in good health now	None	
20 mos.	8 years	Medical report not obtainable. At his work, and apparently in good health	None	Delay in settlement owing to prodigious claim, which was reduced to one eighth. A good deal of exaggeration.
6 mos.	18 mos.	At work again, and not under medical care since settlement. Cannot be traced later	None	Much indiscreet sympathy of friends.
8 mos.	5 years	Has had no illness since, and is in good health	None	
4 mos.	5 years	No illness since, and now in good health	None	Had been in an accident before, and felt "nervous" for two years. Some exaggeration. Medical services dispensed with directly claim was settled.
8 mos.	4½ years	Met him a year after the accident when he said he was quite well. No known illness since	None	A case of considerable exaggeration fostered by leading questions. Exorbitant claim.

No. of case.	Sex.	Age.	Nature of accident.	General outline of case.
16	M.	39	—	See page 235
17	M.	35	Same as 13	Had a blow on the head and lower part of the back. Not stunned. Subsequently had pain in the back and "general nervous shock." Much wilful exaggeration. No objective signs of injury
18	M.	35	—	See page 124
19	M.	58	Bad collision	Stunned momentarily from blow on head. Subsequently suffered from pain and tenderness in dorsal region, with much nervousness and apprehension about himself. Palpitations. Sleepless
20	M.	22	Same as 10	Had a "blow down the whole spine," and became unconscious. Subsequently complained much of pain in the lower part of the back. Had hysterical seizures which began with syncope. The whole family is very delicate
21	M.	28	Same	Said he had a blow on the lower part of the back. Never showed any signs of injury or constitutional disturbance. Professed inability to do anything, and complained of his back. Slight asymmetry of spinal muscles supported the idea of spinal injury
22	M.	45	Most trifling collision	Said he had a blow on his head and was stunned. Pain in the head and sleepless for a short time afterwards. Eight months after the accident, at which time the accident was first heard of, he had, after fatigue at work, an attack of vertigo, followed by some weakness in the left arm
23	M.	42	—	See page 265
24	F., single	32	Very slight collision	Momentarily stunned from blow on face. Subsequent pain in lumbar and dorsal regions, and great hysterical nervousness. A woman of feeble physique. Called a case of "spinal irritation." Has flushings, cold hands and feet, palpitation, pain in back, &c.
25	F., single	28	Same as 19	Had a severe blow on the shin, causing wound which took long to heal. Much nervous alarm. Weak from long confinement to bed. Irregular catamenia

APPENDIX

Date of settlement.	When last heard of after accident.	Condition when last heard of.	Evidences of injury to spinal cord or membranes.	Remarks.
—	—	—	None	
7 mos.	—	Cannot be traced	None	Shortly after he received compensation left the neighbourhood without paying his doctor.
—	—	—	See text	
4 mos.	4¼ years	Fairly recovered. General energy and memory not quite so good. Had no illness since	None	Concussion of the brain.
9 mos.	4 years	Married since accident. Wife dead. One child. Occasionally had "seizures." Now dying of phthisis	None	Frail delicate youth.
6 mos.	6 years	Went to work directly claim was settled. Perfect health since	None	A case of gross exaggeration. A total abstainer and local preacher.
12 mos.	4½ years	In perfect health	None	No evidence of attack being due to the accident, from effects of which he had apparently recovered.
—	—	—	None	
20 mos.	8 years	Looks in good health. Certainly no paralysis. How much neurasthenia it is impossible to say	None	Much harm done by sympathy of friends, who forbad earlier settlement for fear of after-consequences.
5 mos.	4 years	Has not been strong since, but condition thought rather due to constitutional causes than to the accident	None	Comes of a very delicate family, phthisical and strumous, and has herself always been delicate.

No. of case.	Sex.	Age.	Nature of accident.	General outline of case.
26	F.	48	—	See page 130
27	M.	24	Same as 13	Was thrown down, and had a blow on the right lumbar muscles. Shock and vomiting. Later, pain in the back and general nervousness
28	M.	44	No official record of accident	Said he was stunned by a blow on the head, but no evidence of serious injury. Made two long journeys within four weeks of the accident. Subsequently complained of general weakness and utter incapacity for work. Symptoms believed to be due to alcoholism
29	M.	43	Same as 13	Much collapse from blows about face. Severely bruised in various parts of body, especially left shoulder which was squeezed and was followed by tingling of the fingers. He soon recovered from the original shock, and his cuts and bruises did well, but he subsequently suffered much from "nervous shock," loss of memory, and "litigation symptoms"
30	M.	25	Very severe accident at night. One killed on the spot	Was bruised about the legs, and had a deep cut of lip which bled profusely. Some initial shock and slow pulse. At the end of two months said he was quite well
31	M.	25	Same	Much shock. Simple fracture of left femur and comminuted fracture of right tibia. Rallied well. Union good in natural time. No subsequent symptoms of "nervous shock"
32	M.	35	Same	Had a simple fracture of the left femur and considerable collapse from which he soon rallied, and in himself did well. A troublesome cough was supposed to be the cause of mobility, which prevented union. Later, ivory pegging and resection both failed to unite the bone. Leg amputated four years after accident, when he died of pyæmia. Never had any symptoms of general nervous shock

APPENDIX   305

Date of settlement.	When last heard of after accident.	Condition when last heard of.	Evidences of injury to spinal cord or membranes.	Remarks.
—	—	—	See text	
2 mos.	—	Left the neighbourhood, and cannot be traced. Had, however, been at work as usual	None	
2½ years	3½ years	Died. No precise account obtained of last illness. Had resumed work, but abandoned it from "nervous prostration"	None	Had sunk from a good to a very low position. Exorbitant claim many months after accident. No doubt of chronic alcoholism, from which he died.
10 mos.	8 years	Medical report not obtainable. At his work, and apparently in good health	None	Made an enormous claim, which was reduced by litigation to one fifth. There was undoubted exaggeration.
5 mos.	—	Cannot be traced	None	A foreigner.
8 mos.	8 mos.	In good health. Rather lame from shortening	None	A foreigner, and cannot be traced further.
2 years	4 years	Autopsy revealed a "second perfectly united fracture lower down." The fragments at the ununited part were found separated by a considerable layer of muscle, and the surgeon who amputated believed that "these upper fragments could never have been in apposition"	None	

20

No. of case.	Sex.	Age.	Nature of accident.	General outline of case.
33	F., single	25	Same	Had a severe wound on forehead and was stunned. Able to be carried home in three weeks, wounds having well healed. Later had general nervousness and occasional palpitation. Severe mental shock from near connection having been killed beside her
34	M.	47	Collision	Dazed. Had a blow across the shoulders. Subsequently had much pain and stiffness in the loins and back, with tenderness and hyperæsthesia over the dorsal and lumbar vertebræ. In bed seven weeks, so stiff that he had difficulty in moving, and his limbs felt "heavy, numb, and weak." Also said he had to use extra effort to micturate. In three months was moving about again, and presented no objective signs, but complained much of nervousness, &c.
35	M.	28	Same	Was cut about the legs and had a blow on the chest. Slight collapse. Had some pleurisy and consolidation at site of blow. No symptoms of general nervous shock. Improving at end of three months
36	F., single	28	Same	Had a blow across the small of the back, but continued at work until pain compelled her to stop. Then got into a very hysterical state which was nursed by sympathy of friends. Never any objective signs
37	M.	53	—	See page 267
38	M.	24	Bad collision.	Severe collapse. Bad compound fracture of both bones of one leg, and comminuted fracture of both bones of the other. Recovery natural. Never any symptoms of general nervous shock
39	M.	30	A rather sharp collision.	Had a blow on the side of his face, and a cut of one eyelid. Not stunned. Was in bed for three days, crying and hysterical, and had dilated pupils. Resumed work on tenth day
40	M.	29	Same	Blow on face. Stunned. Vomited. In bed for five days, and soon afterwards tried to work, but broke down and had uncontrollable fits of crying. Then went away for a month's rest and came back well. At work afterwards up to time of settlement

APPENDIX 307

Date of settlement.	When last heard of after acciden	Condition when last heard of.	Evidences of injury to spinal cord or membranes.	Remarks.
6 mos.	—	Cannot be traced	None	Was rapidly improving when claim was settled.
5 mos.	4½ years	In excellent health, but is more easily tired by long walking	None	Some exaggeration. Made a very large claim.
5 mos.	4½ years	Never regained the same condition of strength as before. Health variable and precarious	None	A weakly man. Of a phthisical family.
6 mos.	4½ years	Soon resumed work, and now in good health, but is easily fatigued, and suffers from her back when tired	None	
—	—	—	None	
6 mos.	—	Cannot be traced	None	
1 month	5 years	In good health, and at usual work. Generally "more nervous" and does not sleep so well	None	No exaggeration. Very moderate claim.
4 mos.	5 years	Continued good health	None	Moderate claim. No exaggeration.

No. of case.	Sex.	Age.	Nature of accident.	General outline of case.
41	M.	50	Slight collision	Had a severe bruise, which sloughed, on one leg. Slow in healing. In a few days had pain in small of his back, and felt nervous and easily fatigued. This soon passed away
42	M.	30	Sharp collision	Had a cut on one leg. Dazed. In two days had pains all over, and especially in small of the back. Soon mended, and was at work in five weeks
43	M.	43	Same	Had a blow on legs and small of back. Pain in these places for a few days, but never any symptoms of shock, either at time of accident or afterwards
44	F., married	52	Collision	Was thrown down in the carriage and dazed. Very frightened, and in about an hour became hysterical. In bed for a month suffering from pains across the loins and shoulder-blades. Constipation and difficulty in passing water. For four months she had "pins and needles" in the right leg. She continued excessively nervous and hysterical, but there were never any objective signs of central injury. At time of settlement she was well nourished and in good bodily health, but she complained of much stiffness about the loins. A stout heavy woman
45	M.	30	Very slight collision	No evidence of injury, but called in a doctor, and complained of his "spine." Never any signs of illness
46	M.	31	Same as 42	Had a blow on both knees, but did not think himself hurt and went on to work. Soon had to give up, feeling sick, and in twenty-four hours had pain in his back. Had tenderness over last dorsal vertebra. Subsequent nervousness, but no objective signs of ill-health
47	M.	50	Same	Had a contused wound on the cheek, and a blow on his legs. Felt dazed. Had some pain for a few weeks at site of wounds after they were healed. No later nervous symptoms
48	F., single	19	Same	Wound over left eyebrow. Momentarily unconscious. Nervous and hysterical for a few weeks. Early recovery

Date of settlement.	When last heard of after accident.	Condition when last heard of.	Evidences of injury to spinal cord or membranes.	Remarks.
4 mos.	5 years	In perfect health	None	A gentleman in easy circumstances.
5 weeks	—	Cannot be traced	None	
6 weeks	—	Cannot be traced	None	A bricklayer. Left the neighbourhood shortly afterwards. No exaggeration.
15 mos.	5 years	Still has pain in her back, and finds it difficult to stoop or to rise from her chair. Is obliged to use sticks	See outline	May be regarded as a case of permanent injury. Stiffness and incapacity, in all probability, due to severe muscular and ligamentous sprain about hips and loin.
6 weeks	—	A pedlar. Cannot be traced	None	Gross exaggeration.
2 mos.	—	Soon afterwards left the neighbourhood. Not under medical care after claim was settled	None	Some exaggeration.
3 mos.	—	Cannot be traced	None	Moderate claim. Declined the services of a lawyer who offered to "take up his case."
3 mos.	—	Cannot be traced	None	As above. Daughter of No. 47.

No. of case.	Sex.	Age.	Nature of accident.	General outline of case.
49	M.	59	Same	Said he was thrown backwards and forwards, that he was unconscious, and when he came to vomited many times. A fortnight after the accident he showed no signs of injury or symptoms of constitutional disturbance, but he complained of "excruciating agony" in his back and head. Case full of inconsistencies and many lies told. A venerable member of the profession, now no more, thought he had "effusion on the ventricles of the brain." From first to last not a single sign of ill-health
50	M.	20	Same	Had no blow anywhere, but was nervous and sleepless for a fortnight, and was kept at rest for so long when he was perfectly well
51	F., single	28	Same as 39	Had two black eyes and a blow on the wrist. Stunned. In bed a month with weakness and nervousness. Early recovery from nervous symptoms, but long trouble from tenio-synovitis at site of blow on wrist
52	M.	30	Severe collision	Had two ribs broken and a bruise over one ear. Momentarily stunned. Suffered from after-symptoms of concussion of the brain—pain in the head, giddiness, &c. At end of two months was rapidly improving
53	M.	40	Very severe collision	Much bruised about forehead, occiput, face, shoulders, and lower dorsal vertebræ. "Light-headed" for a few days. Gradually recovered and resumed work
54	F., married	27	Same	Much frightened. Had bruises on shins and left shoulder. In bed for a few days, but later there were no symptoms, though she worried about compensation. Was somewhat hysterical, and complained of her back
55	M.	45	Same	Cut about face and head. Never rallied from collapse, and died within twenty-four hours. Ribs found broken and pelvis smashed in three places, with extensive hæmorrhage in pelvic cavity
56	M.	41	—	See page 230

Date of settlement.	When last heard of after accident.	Condition when last heard of.	Evidences of injury to spinal cord or membranes.	Remarks.
7 mos.	4 years	Was then in a second accident, and attempted the same performance again. Had gone to work directly his claim was settled, and been in ordinary good health since	None	A gross case of exaggeration, supported by "leading" examination.
6 mos.	—	Cannot be traced	None	Cause of delay in settlement cannot be discovered.
10 mos.	—	A servant. Cannot be traced	None	
3 mos.	7½ years	At his usual work, and in good health	None	
4 mos.	4 years	Three years after accident had severe scalp wound from fall. Had good health continuously since, but says he has "never felt quite the same"	None	A moderate claim. No exaggeration. Concussion of brain.
3 mos.	7 years	Resumed work soon after settlement	None	
—	—	—	—	
—	—	—	None	

No. of case.	Sex.	Age.	Nature of accident.	General outline of case.
57	M.	52	Slight collision	Some shock. Contusions about right shoulder. Never made any subsequent attempt to use his arm. As time went on symptoms of "general nervous shock," due largely to staying in-doors and agitation about his claim, which was very large. Much exaggeration and some thought wilful imposture
58	M.	49	Very bad collision	Was dazed and felt very sick. No known blow anywhere. Soon felt sore all over, and was sleepless for several nights. Suffered much from pain in head and neck. Became very irritable and emotional, with tendency to cry. Seemed much aged by the accident, and for a time looked thin and worn and ill
59	M.	39	Collision	Was bruised on the back by a falling portmanteau and felt shaken. Returned home next day, and seemed to be suffering from "shock and bruises." Said in a few days to have had an attack of violent delirium in which he had to be held down. Recurrence of this attack after an interval of three weeks, during which time he had slept heavily. Most violent, and threatening suicide. Never any disturbance of pulse or temperature, nor any paralytic symptoms. Later "attacks" came on whenever he was visited, but it was proved that in the intervals he lived like a person in ordinary health, though not going out of doors
60	M.	56	Terribly bad collision	Had a severe scalp wound and exfoliation of large portion of outer table of skull. Wound took six months to heal. In good health at end of two years, but memory not quite so good
61	F., married	55	Same	Had a bruise on the lower part of the back. Does not appear to have suffered much at the time. A few weeks later had a severe attack of pneumonia. At end of two years in good health, but very nervous and most apprehensive as to the prominence, quite natural, of the 7th cervical vertebra

Date of settlement.	When last heard of after accident.	Condition when last heard of.	Evidences of injury to spinal cord or membranes.	Remarks.
12 mos.	—	Soon after settlement left the neighbourhood and cannot be traced	None	Litigation.
6 mos.	4 years	After settlement went abroad. For long looked wretchedly ill, but gradually recovered his ordinary appearance, and now does his usual work. Has had no illness since	None	Moderate claim. Easy circumstances. No exaggeration.
14 mos.	8¾ years	Medical report not obtainable. Alive and well. Certainly has no paralysis of any kind, but is at times "very nervous"	None	A very exorbitant claim. Litigation. However real the early condition may have been it is tolerably certain that the later manifestations of neurotic disturbance were largely, if not altogether, under his control. In the attacks seen he used to lie down, say he was "going off," and decline to speak any more. He never hurt himself, and never really fainted. An emotional undercurrent of gross exaggeration for purposes of claim.
2 years	2 years	Gone abroad and cannot be traced later	None	Concussion of brain. Settlement delayed by special circumstances.
2 years	2 years	As above	None	As above.

No. of case.	Sex.	Age.	Nature of accident.	General outline of case.
62	M.	75	Same as 42	Was bruised on the forehead and on the lower dorsal and lumbar region. Stunned. In bed for a few days. Later complained of pain in the head, stiffness of the back, loss of memory. At end of two months no symptoms of ill-health
63	M.	20	Same as 53	Had no blow except on one arm, and went on to his work. Vomited two hours after and felt shaken. Later had pain in head and neck and was sleepless
64	M.	44	Same as 15	Had no bodily injury but was "shaken." Later he complained of headache, loss of memory, inability to apply himself. Became irritable and bad-tempered. Always somewhat eccentric
65	M.	28	Very trifling collision	Portmanteau fell on his head and momentarily stunned him. Soon felt shaken and broke down at work. Had headache, pain in the neck, flushing of face. Was at first very sleepy. After six weeks' complete rest he gradually improved. Eight years before had been laid up with "general nervousness from over-work at college"
66	M.	24	—	See page 264
67	F., married	26	Bad collision	Dazed. Soon had pain and tenderness over the lower dorsal vertebræ. Pregnant. Much fear of miscarriage, having had miscarriage once before at seventh month from fright. Course of pregnancy and labour not interfered with. Subsequently much emotional disturbance, coupled with undoubted exaggeration
68	M.	53	Slight collision	Blow on back of head and bruise of left shoulder. Felt no ill effects until he got home three days afterwards. Then had pain in the arm, oppression of the head, and aching at the lower part of the back. His left arm felt numb and stiff. An unhealthy plethoric man. A free liver

Date of settlement.	When last heard of after accident.	Condition when last heard of.	Evidences of injury to spinal cord or membranes.	Remarks.
4 mos.	5½ years	Died 5½ years after accident of "old age." Had been in good health and able to go about until within a few weeks of death	None	Much exaggeration and exorbitant claim. Had fallen into very poor circumstances.
6 mos.	—	Left the neighbourhood. Known, however, to have been perfectly well and at work again	None	Delay in settlement owing to large claim. Some exaggeration.
8 mos.	5 years	At usual work, and in as good health as before the accident	None	Made an exorbitant claim. Much exaggeration.
3 mos.	—	Cannot be traced	None	Concussion of brain in a nervous subject. Moderate claim.
—	—	—	None	
9 mos.	3¾ years	Says never been so well since accident. Very nervous at times. Pain in the back, cold limbs, and deadness, &c. Looks thin and careworn; constantly fretting. Had three miscarriages since accident in addition to the child born after it. A question how far her state may be due to the exhaustion of frequent gestation	None	A most outrageous claim. Litigation. Jury awarded one sixth of the amount claimed. There is little doubt that the condition of this patient was rather an outcome of the worry attendant on expected compensation than on the real shock originally sustained.
3 mos.	3½ years	No medical attendance after settlement. Had no illness since except indigestion	None	Exaggeration. Large claim.

APPENDIX

Sex.	Age.	Nature of accident.	General outline of case.
M.	52	Same as 67	Severe shock from blow on head. Both eyes black. Pulse 52 for three days. Gradual recovery, and in nine weeks went to work, but felt very unfit for it. Later, very nervous and worried about his business. Much worry also about his claim. Increased presbyopia led to examination of eyes, when cataract of both lenses was found, and on this a very large claim was based. There were no other ocular changes, and one distinguished surgeon whom he consulted, but whose opinion was not mentioned until long afterwards, could discover no connection between the accident and the cataract
M.	39	Sharp collision	Blow on shin. Momentarily stunned from blow on forehead. In bed fourteen days. Suffered afterwards from headache and palpitation, and was generally nervous. To work in six weeks, and was then daily improving. Looked aged and worn
M.	58	Very slight collision	The only man hurt. Bruise on side of chest. Away from work a month. Atheromatous aorta
M.	36	Same as 60	Stunned from blow on head. Away from work nearly three months. Seven months after accident was still complaining of pain in the head, of being easily fatigued and startled, and of bad sleep. Manner very agitated. Pulse weak. At his work long before claim was arranged
F., single	29	Same as 10	Had no blow anywhere, but was very much frightened. Two days afterwards began to have pain in small of back, and became very hysterical, screaming at the least noise. Never any alteration in pulse or temperature. Began to improve in six weeks
M.	38	Same as 8	Blow on side of head and bruise on lower part of back. Indoors for fourteen days, suffering from pain along the whole vertebral column and tenderness about the coccyx. Also extremely nervous and hysterical. At end of six months was partially at work, but looked haggard. He complained of nervousness, loss of memory, loss of sexual power, and creeping sensation in his legs. At the end of thirteen months there were no objective signs of central injury

APPENDIX 317

Date of settlement.	When last heard of after accident.	Condition when last heard of.	Evidences of injury to spinal cord or membranes.	Remarks.
14 mos.	6 years	In good health, and at his usual work. Said to be more irritable than he was. No report about cataracts	None	Concussion of brain. Delay in settlement from large claim.
3 mos.	—	Left the neighbourhood and cannot be traced	None	Concussion of brain. No exaggeration.
4 mos.	10 mos.	Still at work and in good health. Cannot be traced further	None	Small claim. No exaggeration.
2 years	—	Cannot be traced	None	Concussion of brain. Special circumstances delayed settlement.
6 mos.	5 years	Good health. Married, and had one child. Certainly been very nervous since, starting, for instance, at the closing of a door	None	Sheer fright.
13 mos.	8½ years	Occasionally has slight asthma, but is otherwise in excellent health. Thinks his sexual desire is not what it ought to be in a man of his physique, but he gives no standard for comparison	See outline	Gouty diathesis.

318   APPENDIX

No. of case.	Sex.	Age.	Nature of accident.	General outline of case.
75	M.	30	Same as 23, and another very slight collision two days before	Said that in the first accident he had a blow on his left side, and that he suffered from fright and nervousness, but did not leave his work. In the second he had a graze on one shin and again was terribly frightened. Then left work, and said he was laid up for six weeks with "sleeplessness, loss of appetite, inability to enjoy anything, and complete nervous prostration." At end of four months no sign of ill-health whatever, but he complained of impaired memory, mental dulness, and inability to occupy himself
76	M.	39	Same as 58	Stunned from blow on back of head. Laid up four days. Soon mended in himself, but a high degree of hypermetropia was revealed, and twelve months after accident he had convergent strabismus. This was partially rectified by lenses. No evidence of structural damage to eyes. Improved under treatment of a well-known ophthalmic surgeon, but at time of settlement some strabismus still remained. A strong muscular man
77	M.	55	Bad collision	Momentarily dazed, but had no blow. Soon felt tremulous and giddy, and suffered for a few days from sleeplessness and irritability. After some weeks at the seaside, and ten weeks after the accident, he felt well and ready for work, but did not return to it, thinking it better to wait. He then again began to be nervous, sleepless, and tremulous, and had one attack not unlike delirium tremens. Later he suffered from much mental depression, but showed no signs of bodily ailment, simply professed inability to work. Had been a somewhat free liver
78	M.	25	Very slight collision	Admittedly had no known injury, but next day felt shaken. Then had pain in small of back and took to bed, where he lay for nine weeks, his back being stiff and painful. During this time lost appetite and strength, and became nervous and emotional. No signs of disease at date of settlement
79	F., widow	45	Slight collision in shunting	Stunned by blow on head. Vomited. Had been out of health for some time, and was on way to country for change of air. The doctor thought her recovered from effects of accident in a fortnight

ate of ttle-ment.	When last heard of after accident.	Condition when last heard of.	Evidences of injury to spinal cord or membranes.	Remarks.
6 mos.	—	Cannot be traced	None	Exorbitant claim. Gross exaggeration.
16 mos.	7 years	At his usual work ever since, and in good health	None	A very unusual and remarkable case.
8 mos.	4½ years	More susceptible to worry and excitement. Frequently under treatment for indigestion. Suffered also from much melancholy. Is at work, but feels prostrated by railway travelling	None	Temperament predisposed him to suffer much. Made a very large claim, which admittedly worried him.
11 mos.	—	Cannot be traced.	None	Delay in settlement due to largeness of claim. Much exaggeration.
4 mos.	—	Left the neighbourhood and cannot be traced	None	Made a large claim on basis of ill-health which really preceded accident.

No. of case.	Sex.	Age.	Nature of accident.	General outline of case.
80	M.	37	Same as 77	Blow on abdomen, and was much shaken. Vomited. Suffered from indigestion and constant vomiting for six weeks. Much reduced in strength and weight, and became very nervous and hypochondriacal. Five months after the vomiting ceased he was complaining of pain and weariness in the back as if he had sprained himself, and of feeling prostrate and good for nothing
81	M.	49	Same as 60	Severe blow on head. Unconscious fifteen minutes. Had wound of hand and extensive laceration of leg which kept him in bed five weeks. Was under medical care for three months longer, and then resumed business
82	M.	24	Same	Stunned by blow on parietal region. Went home alone in five days, and then lay for four weeks in a state of extreme prostration, in which he never spoke except when going to be fed. It was thought he might die. Was away from work about a year, during which he was weak and upset, and felt very fatigued in the back, but had no objective signs of central injury. At end of two years he looked in good health, but made the most of his ailments when compensation questions then arose
83	M.	56	Same	Unconscious. Vomited. Severe bruise of left arm, and subsequently some periostitis of humerus. Did not lie up, although feeling very nervous and shaky. In a few weeks resumed business, but became more nervous and weak, and had to go away for some months' rest
84	M.	56	Same as 67	Bruised on one shoulder and "all along the back." Had his lip cut. Felt "numbed" all over. No signs of shock. For a week the left leg was colder than the right, and it perspired more than natural. Later he complained much of pain when stooping in his back. Four months afterwards, when to all appearances well, he professed inability to do anything
85	M.	52	Same	Slightly shaken and frightened, and had a bruise on forehead and one leg. No after-symptoms

Date of settlement.	When last heard of after accident.	Condition when last heard of.	Evidences of injury to spinal cord or membranes.	Remarks.
12 mos.	3½ years	Never felt quite so vigorous or capable of exertion. Digestion very uncertain. No objective signs of disease. Following his business as usual	None	Made a very exorbitant claim.
2 years	2 years	In good health, but knee still rather stiff	None	Settlement delayed by special circumstances. It had no influence on course of illness.
2 years	5½ years	No special report obtainable, but he has since married and begotten a child	None	As above re settlement. Genuine case of extreme nervous exhaustion.
2 years	2 years	Not doing so much work as before, but from age and position was entitled to rest. Had to give up shooting	None	As above re settlement. A man of nervous temperament.
11 mos.	3½ years	Looks well and says he is so	See outline	In pecuniary difficulties. Exorbitant claim. Much exaggeration.
8 mos.	—	Cannot be traced	None	Made a very large claim. Delay in consequence.

No. of case.	Sex.	Age.	Nature of accident.	General outline of case.
86	M.	58	Same	Dazed from great fright. Had a small wound and bruise on one arm. Soon complained of pain all over, especially in the back. Later became very nervous, and admittedly, from his previous experience, was anxiously on the look-out for symptoms
87	M.	—	Same	See page 263
88	F., widow	50	Same	Blow on back of head. Stunned. Much bruised in many parts, but worst of all over the left hip and pelvis. Subsequently had pain in the back, and much stiffness about hip and thigh, which were very painful on movement. Complicated by rheumatoid arthritis. Pseudo-palsy. In other respects she improved, and at time of trial was in herself pretty well
89	M.	53	Same	Much dazed. Bruises on the back and both shoulders. Subsequently much pain in whole vertebral column and stiffness at the lower part. General nervous shock. Cold feet, sweating, much nervousness. Nothing would induce him to leave his room for several months. Never any objective signs, but he used to complain that his legs occasionally felt " numb."
90	F., married	36	Same	Under treatment at time of accident for prolapsus uteri. Considerably bruised about hips and loins. Unconscious. Vomited. Suffered much from pain and hyperæsthesia of back which lasted for many months. As time went on became more and more nervous, felt prostrate and good for nothing. Much of this was due to worry about compensation

APPENDIX

Date of settlement.	When last heard of after accident.	Condition when last heard of.	Evidences of injury to spinal cord or membranes.	Remarks.
10 mos.	—	Lives out of the country and cannot be traced	None	Had been in a bad accident 3 years before, and was laid up for 11 months with general nervousness. From that, however, he had perfectly recovered.
—	—	—	None	
5 mos.	—	Cannot be traced. Doubtless still suffering	None	Deserving of much sympathy. Humble station in life. Friends foolishly allowed her case to be taken up by a solicitor who went round directly after the accident offering his services. Trial of course. Verdict for less than the company had offered. It was subsequently learned that the whole of the money was kept by the lawyer, this poor woman not getting one single farthing.
10 mos.	3½ years	After settlement soon began to go about and much improved. Still has some stiffness in loins, and can't lift heavy weights so well nor walk so far	None. See outline	Made a very large claim. Unconscious exaggeration.
18 mos.	3¾ years	Certainly not the same strong woman she was. Has had no illness since, but every now and then has pain in the small of the back	None	Note previous history. Exorbitant claim.

No. of case.	Sex.	Age.	Nature of accident.	General outline of case.
91	F., married	38	Same	Dazed. Struck between the shoulders and soon began to feel sore all over. Long journey home in a few days. Then began to suffer more from pain in small of back, and complained also of "numbness" over the whole left side, limbs, trunk, head, and neck. No real anæsthesia. Hyperæsthesia of back. Great hysterical nervousness about herself, and fear of future ill consequences
92	M.	36	Same	Much alarmed and dazed. Vomited profusely. Felt much shaken. Bruised on left ribs and both elbows, and later tingling in course of left ulnar nerve. In six weeks attempted work, but on failing to continue at it became terribly depressed, as his employers then dismissed him. Improved after change of air. Much asthenopia, which was ultimately relieved by convex glasses. Never any objective signs except those due to blow on elbow. Suffered also from much pain in the back, and from symptoms of general nervous shock
93	M.	32	Same	Unconscious. Blow on knee and right side of neck. Much collapse. Pulse 48 for twenty-four hours. Left pupil during first day said to have been dilated and sluggish. Pain across sacrum. Went away for rest and change at end of ten days
94	M.	29	Same as 60	History of having been stunned from severe blow on head. Unconscious for four or five hours. In bed three weeks. Suffered from pain in the small of the back. In four months returned partially to work, but suffered thereafter, in a gradually lessening degree, from pain in the head, occasional giddiness, and inability to work as long as before. After concussion symptoms of Hutchinson
95	M.	29	—	See page 133
96	M.	38	Same as 77	Stunned by blow on nose. Found himself on floor of carriage. Complained of general nervousness and constant pain between shoulders, made worse by exercise or stooping; of a sense of "confusion in his head" when many people were near him, and of "numbness" below the knees which came on after walking. Never any objective signs

Date of settlement.	When last heard of after accident.	Condition when last heard of.	Evidences of injury to spinal cord or membranes.	Remarks.
18 mos.	7 years	In as good health as ever she was	None	Symptoms kept up by family trouble. Much genuine nervousness. Had for years suffered from uterine derangement and poor digestion.
15 mos.	6 years	Now manager to a large house of business. Is very well, never having felt any bad after-effects from the accident	None	Symptoms aggravated by ungenerous conduct of employers, and also by alarm from erroneous opinion about his eyes, a small patch of choroidal pigment at the margin of one disc having been diagnosed as a "retinal hæmorrhage."
8 mos.	7 years	Has had good health. Attending daily to a large and important business	None	Made a very exorbitant claim.
2 years	2 years	Lives abroad and cannot be further traced	None	Concussion of brain. Settlement delayed by special circumstances.
—	—	—	See text	—
12 mos.	3 years	Ailed for some time after his claim was settled, but now perfectly well	None. See outline	A man of highly nervous temperament.

APPENDIX

Sex.	Age.	Nature of accident.	General outline of case.
F., married	35	Very trifling collision	Bruise on one elbow Much frightened. Became very nervous. Hysterical attacks. Suspicion of alcoholism. Six months after accident had a miscarriage. Very weak and nervous after it, and two years after accident had "functional paraplegia" which lasted for about five months. Circulation very feeble. Feet and hands always cold. Did not waste
M.	30	Same as 68	Fat, unwieldy man. Slight blow on back of head, right shoulder, and left knee. Momentarily unconscious. Did usual business for two days, then made a long journey and broke down. Suffered from headache and sleeplessness. Generally nervous and good for nothing. At end of two months much better, and then began to exaggerate
M.	33	Slight collision	Momentarily unconscious from blow on head. Vomited. In bed nine weeks, feeling utterly weak and prostrate. Also very hysterical. For several months was in a very nervous state. Had headache, sleeplessness, heat of head, and weakness of right leg. No objective signs, but he looked heavy and listless
M.	59	Collision	Momentarily unconscious from blow on the head. In bed nine weeks, feeling stiff and sore all over. At end of that time nothing whatever could be found the matter with him. Two years after accident said he could not do any work because of pain in his head and back. The bodily functions were all natural, and he did not then appear ill
F., married	51	Same	Blow over one eye and momentarily unconscious. In bed eight weeks, suffering much from prostration, and was stiff and painful all over, especially in her back. At end of two years was still complaining, looked feeble and ill, and was extremely nervous about herself

Date of settlement.	When last heard of after accident.	Condition when last heard of.	Evidences of injury to spinal cord or membranes.	Remarks.
3 years	5 years	Has become very stout, but is now in excellent health. No paralysis whatever	None	Compensation had no influence on history.
13 mos.	3½ years	Not under medical care since settlement. Believed not to have suffered any permanent injury	None	Delay owing to exorbitant claim. Pecuniary difficulties.
10 mos.	13 mos.	Greatly better 3 months after claim was settled. Headache gone, he was less nervous, and he felt nothing wrong with his leg. Cannot be traced further	None. See outline	Concussion of brain. Made a very large claim to the worry about which his symptoms were thought largely due.
4½ years	5½ years	Still unable to work	None	Very large claim. Man declined to settle sooner, preferring to wait and see what would happen. In good circumstances, and inability to work due to fact that he had small occasion to work.
4½ years	5½ years	Now crawling about with a stick, and it is believed she will never be worth anything again	None	At the menopause.

No. of case.	Sex.	Age.	Nature of accident.	General outline of
102	M.	42	Same as 60	Cut and bruised about face an(legs. Laid up fourteen da) other fortnight resumed bu: soon. Continued at it partia and then went to seaside. "fainting-fit," and says he ! side. Laid up for fourteen (three months more had el: worse attack. Then to bu: end of twelve months a thir(came very nervous—"grie pacity." Two years after nervous, could not bear tl complained of "weakness' "numbness" in the arm an tive signs
103	F., married	43	Same	Severe bruises on both knees. Though able to walk abou not attend to household duti Later much exaggeration
104	M.	38	Same	Cut and bruised about head an(up partially for about eight v that, but at end of two years tainly an exaggerated way) heaviness of the head, and (No objective signs. Endea that he had been ruptured, b: could not be found it was that it had been cured by h
105	M.	50	Rather sharp collision	Unconscious. Vomited. Blov In bed five days, having mu between the shoulders. La and stiffness in the lumbo-s; weak and shaky. Off work
106	F., widow	78	Sharp collision	Severe blow on forehead and In bed for some weeks. Ra: and at the end of seven m ptoms. Weak, however, an work. Suffers much from later had an attack of jaun(
107	F., married	32	Slight collision	Frightened and shaken. Mome: In bed three weeks, sufferin; back. In five weeks went t(

APPENDIX 329

Date of settlement.	When last heard of after accident.	Condition when last heard of.	Evidences of injury to spinal cord or membranes.	Remarks.
2¼ years	8 years	In perfect health and at his usual business	None	Very gross exaggeration. Outrageous claim based on untrustworthy commercial losses. Litigation. Not a straightforward case. Compensation was delayed by special circumstances, and when it arose he unquestionably nursed himself into a state of illness.
2¼ years	8 years	To all appearances in very good health	None	
3 years	—	Not able to hear of him since	None	Compensation delayed by special circumstances. Much exaggeration, conscious and unconscious.
3 mos.	10 mos.	Lost sight of. Lost his place from drinking	None	An "habitual drunkard."
10 mos.	3 years	Continued to suffer much from headache, and died of "old age" æt. 81, 3 years after the accident.	None	Concussion of brain. Age.
1 year	7 years	In good health, but is easily fatigued, and after much exertion has	None. See outline	Made a very large claim.

No. of case.	Sex.	Age.	Nature of accident.	General outline of case.
				took daily exercise, and came home greatly better. Then resumed her household work, but compensation also at the same time began to engage attention, and she complained more of pain—hyperæsthesia—in her back and head, general weakness, sleeplessness, poor appetite, and "numbness in the right leg after walking." No objective signs. Catamenia regular
108	F., married	50	Very slight collision	Slight blow on head. Dazed. In bed eight days with pain in back and head. After change of air returned home, but found herself unable to do anything, and at end of nine months after accident was complaining terribly of her back and head. Former extremely hyperæsthetic. Her back had been fearfully punished by huge blisters. Very nervous and apprehensive. Highly emotional
109	F., married	53	Same as 68	Much bruised about right arm and leg, which for a time felt numb and heavy. In bed ten weeks. Rallied as far as her strength would allow, and was able to get into country six months after accident. Fourteen years before had acute rheumatism. Now has mitral regurgitation and great cardiac irregularity. Shortness of breath and palpitation induced by accident
110	M.	44	—	See page 172
111	M.	27	Sharp collision	Momentarily stunned from blow on head. Did not think he had been much hurt, but in two days felt so shaken that he had to lie up. Had tenderness at nape of neck and oppression of the head. Hyperæsthesia of the scalp. Began to mend in a month
112	M.	48	Rather sharp collision at night	A man looking much older than his years. No marks or bruises. At work for two days, and then felt shaken and sore. Had pain in back and head, and especially about the sacrum when he stooped. Sensation also of cold water running up and down his back. Thought well enough to resume work in four weeks, but then made large claim, began to worry about compensation, stay at home, and undoubtedly exaggerate

Then last heard of after accident.	Condition when last heard of.	Evidences of injury to spinal cord or membranes.	Remarks.
	a return of pain in the lumbar region. Catamenia regular. Not pregnant since		
4 years	Nervous for another 12 months, and then improved. Now in good health. Back feels weak after much exertion. Does not work so actively as before	None	At the menopause. Remedy worse than the disease. Claim and treatment based on supposed disease of cord, but there was never the least sign of it. In good circumstances.
4½ years	After continued suffering from the heart disease, died 4½ years after the accident	None	Previous disease.
—	—	None	
—	Left the neighbourhood and cannot be traced	None	Moderate claim. No exaggeration.
3 years	Very soon at work after settlement. Has had good health since and been able to knock about as usual	None	Exaggeration.

No. of case.	Sex.	Age.	Nature of accident.	General outline of case.
113	—	—	—	Same as 23. See page 265
114	M.	45	Slight collision	Had a small wound on one knee and a slight blow on sternum. Did not complain much for the first fortnight, and then took to his bed for two months. Five months afterwards was complaining of great nervousness and inability even to attempt to do anything. Also said he had profuse hæmoptysis, blood coming, he was sure, from the part of the chest where he had been struck. No doctor ever saw this, and what blood there was came from very congested fauces. No symptoms whatever of disease
115	M.	24	Same as 77	A dwarfish, ill-formed, half-witted man. Seven in a family of eleven had died in infancy. Blow on head. Laid up about three weeks. Since that time said to have been incapable of work, complaining of pain in the back and inability to do anything. Eighteen months after accident there was no sign of ill-health
116	M.	29	Same as 60	A frail delicate man, who stammers. Left clavicle broken, and he was bruised on the face, left side, arm, and leg. Not stunned. Had a small flesh wound in lumbar region. Not confined to bed. After rest in country returned home in three weeks, when he seemed in good health and had nothing the matter with him except the broken collar-bone. Two months after accident married, and two months later still came under medical care for sense of debility and inability to work without fatigue. When seen thirty months after the accident he was complaining of great nervousness, inability to concentrate attention, anxiety about himself, restlessness, sleeplessness. He had a very frequent pulse and stammered much. No sign of anything like paralysis
117	F., married	38	Same as 68	Much bruised about right hip, sacro-iliac region, and shoulders. Laid up several weeks. Later great stiffness about shoulder and hip, latter more especially, and she had much pain and tenderness about sacrum and coccyx, so that she dreaded to walk. Hypcræsthesia of back. Highly nervous

Condition when last heard of.	Evidences of injury to spinal cord or membranes.	Remarks.
—	None	
Had long been at work and was in perfect health	None	Gross exaggeration. Almost a malingerer. Litigation avoided.
Cannot be traced further	None	Much exaggeration by friends, who made exorbitant claim on his behalf. Inquiry failed to elicit any change in mental state as result of the accident.
Cannot be traced	None	Compensation delayed by special circumstances. An apparent exception to the usual rule about nervous shock after fractures, but the history shows that the later condition must have been due to other causes than injuries received. He had recovered, and only after several months became ill. He had been unsuccessful in business. Made a very large claim. Symptoms very largely due to compensation worries.
Left the neighbourhood shortly afterwards and cannot be traced. Wife of a labouring man	None	Large claim, but no conscious exaggeration.

No. of case.	Sex.	Age.	Nature of accident.	General outline of case.
118	F., married	32	Slight collision at night	Could give no account of the accident, but had gone on her journey and then walked from the station home. Later complained of pain in the back. Pregnant. Labour natural four months after accident. After recovery therefrom still complained of her back and of tenderness about the coccyx, and said she could not attend to her houschold work. Was able, however, to enjoy all the festivities of a fashionable watering-place
119	F., married	38	Careless shunting in station	Had a severe blow on side of face and head, which caused some periostitis about malar bone; also rather deaf in ear of same side. Nervous and much upset
120	M.	47	Severe collision at night	Asleep. No knowledge of injury. Helped others. Next day confused and could not do business. On third day long journey home. Then had much pain in back and at nape of neck. Pulse 40. Deferred shock. Very nervous and much difficulty in finding right words to express himself. Began work again in five weeks. In two months much better. Pulse 56. Later felt very easily fatigued by work, and became occasionally quite prostrate, and had pain in the head. After a sea voyage well. Pulse then 72
121	M.	36	—	See page 238
122	M.	37	—	See page 114
123	F., single	23	Same as 120	Had no blow, but felt faint and sick. For three days felt stiff about the neck. Became nervous, afraid to travel or go out alone. Some exaggeration on the part of her friends when seeking compensation. Was at work some time before settlement
124	F., married	30	Sharp collision	Very frightened, but not unconscious. Severe bruises on left hip and sacro-iliac region and one arm; also felt her neck twisted. In bed three weeks, suffering from pain in neck and head. In four weeks more, able to be moved to seaside. Five months after accident still complained much of pain in small of back and tenderness at coccyx. Also much stiffness of left hip. Pseudo-palsy

APPENDIX

Date of settlement.	When last heard of after accident.	Condition when last heard of.	Evidences of injury to spinal cord or membranes.	Remarks.
18 mos.	4½ years	Another child since. In good health. In no way affected by the accident	None	Delay from largeness of claim. Much exaggeration of the evangelical type.
5 mos.	3¼ years	Continued very nervous for more than a year. Now quite well	None	No exaggeration.
18 mos.	4½ years	Been in perfect health since, but has always felt a difficulty with figures and accounts of his business	None	A gouty man. No exaggeration. Predisposed to suffer. Travelling under strong presentiment of coming accident. In good circumstances. Sleep had not protected him.
—	—	—	None	
—	—	—	None	
4 mos.	—	Cannot be traced	None	
11 mos.	5¼ years	Now walks well and is in excellent health	None	Litigation by lawyers. No conscious exaggeration.

No. of case.	Sex.	Age.	Nature of accident.	General outline of case.
125	M.	31	Same	Blow on head and much dazed. Slow pulse, 48 for two days; also polyuria. Not confined to bed. Neck stiff for a few days. In three months was much better and went to the seaside, but caught a very severe cold which threw him back, and six months after accident, when claim began to be discussed, was worse, complaining of nervousness, trembling, loss of sleep, no appetite, and occasional weakness in the left leg after walking. No objective signs
126	M.	35	Two very slight collisions. In the first he was the only person who complained. Of the second accident there is no official record	In first accident no knowledge of injury, although he had a "punch on the left ribs." Continued at work for several days, and then complained of pain in the back and head, which kept him an invalid for five weeks. No medical attendance. After being at work for a few days was in an alleged second accident. Again had no knowledge of being hurt, but soon complained that he hardly knew what he was doing, and took to bed. Said he was in bed about seven weeks, complaining of his back and head, and that he used to go dead all over. Did not resume work. At end of twelve months no evidence of injury, but he was full of complaints. Condition believed to be due to chronic alcoholism. Much harm done by leading examinations
127	M.	25	Same as 67	Four years before had been laid up with "nervous debility," as the result of mental shock. No knowledge of injury, but next day had pain in back; also felt faint and weak. After exertion had acute pain in small of back, which laid him up for a week. Improved again, but ten months after accident still had much pain in mid-dorsal region, was sleepless, nervous, and very anxious about himself

APPENDIX

Date of settlement.	When last heard of after accident.	Condition when last heard of.	Evidences of injury to spinal cord or membranes.	Remarks.
11 mos.	5½ years	Appears to be in excellent health	None. See outline	Very large claim. Litigation. Much exaggeration.
13 mos.	18 mos.	Said never to have recovered. Died in uræmic coma. No autopsy. Subsequently learned that diagnosis of alcoholism was perfectly correct	None	Exorbitant claim. Litigation. A case of gross imposture. The late Dr Ogden Fletcher has recorded a similar case. 'Railways in their medical aspect.' Cornish Bros., 1867, p. 73. A male, said to have died twelve months after " from the effects of the railway accident." " On inquiring minutely into this I found that during the last year of his life he had frequent attacks of delirium tremens (from which he had suffered before the accident), and the death was registered as—primary cause 'delirium tremens,' secondary 'cerebral effusion.'"
18 mos.	3 years	At work as usual and very much better than he was. In cold foggy weather has pain in small of back, but otherwise feels fairly well	None	A perfectly genuine case of nervous disturbance in a highly nervous man. Delay owing to his apprehension of the future.

No. of case.	Sex.	Age.	Nature of accident.	General outline of case.
128	M.	35	Rather sharp collision	Dazed and vomited. Said to have had left shoulder dislocated. In bed three weeks. Subsequently he complained of much pain in the shoulder and in the sacral region, of being nervous and easily startled, and of being afraid to travel
129	M.	40	Same	Could not say what had happened to him except that he had been "knocked about terribly." Had a contused laceration on shin which was long in healing. Iodide of potassium necessary in treatment. Has been subject to "rheumatic gout," and a few weeks after accident had an attack of "iritis." No exaggeration. At end of five months was nearly well and partly at work again
130	F., married	44	Carriage off line at night	Much frightened. Shaken. Bruised on ribs and on back of head had severe blow. Wound of hand. Profuse menorrhagia almost directly after accident. This lasted three weeks and much reduced her. Headache, nervousness, sleeplessness, loss of appetite. Able to go home two months after accident
131	M.	27	Same as 124	Jumped from the train and received very serious injuries. Ptosis, strabismus, slight facial paralysis, deviation of tongue, weakness of one leg, and great emotional disturbance. Details of case too long and intricate to give in detail. Probably some injury about base of brain, &c.
132	M.	52	Very sharp shunting collision	His carriage received the full force of collision. No blow, but almost instantly had pain about left side and felt faint and sick. Laid up for a few days after he got home, having much pain in his back and head. A fortnight afterwards was up and about, but was strangely tremulous and nervous. Had much pain and stiffness in small of back, aggravated by movement; also constipation. Chief complaint was weakness and prostration
133	F., single	36	Collision at night	Much bruised about eyes and nose. Considerable shock. Had much pain and stiffness in small of back. Later suffered severely from symptoms of general nervous shock

APPENDIX

Date of settlement.	When last heard of after accident.	Condition when last heard of.	Evidences of injury to spinal cord or membranes.	Remarks.
13 mos.	5 years	At work and in perfect health.	None	Delay owing to largeness of claim. Undoubted exaggeration.
8 mos.	5½ years	Has had no return of iritis. In robust health now	None	Previous ill-health. ? Syphilis.
2 mos.	—	Cannot be traced. Home in remote village in Scotland	None	Concussion of brain.
10 mos.	3½ years	Much better than he was. Objective signs less obvious. Has gained weight and strength. Able to work	See outline	Teaches the wisdom of sitting still rather than of jumping from the train when a collision is expected. The only person permanently injured.
3 mos.	5 years	Perfectly recovered, and at usual work. Felt general nervous disturbance for some time. Has had to give up riding because such exertion made his back ache, but backache is not uncommon in his profession	None	No exaggeration. Very moderate claim.
6 mos.	6 years	Has since married, and enjoyed good health	None	Moderate claim. Concussion of brain, &c.

No. of case.	Sex.	Age.	Nature of accident.	General outline of case.
134	F., married	45	Same	Severe blow and wound on right frontal eminence. Alarming collapse. Moved home in eighteen hours. Laid up in bed five weeks suffering much from head and back. Two months after accident was extremely weak, and complained of pain and numbness at the back of the head, of great nervousness, and of being easily startled. The lower part of her back also felt "numb," and was made worse by movement. Hyperæsthesia in lumbo-sacral region. After walking her legs felt "numb," but there was no anæsthesia, and all movements were perfect. After change of air was improving when claim was settled
135	M.	54	Same	Had a blow on left lower ribs, and a severe wound on the shin. This took long to heal and kept him on the couch six weeks. Cardiac irregularity. Atheroma. Improved at end of three months
136	M.	34	Same	Seemed shaken and did not continue journey. Much pain about lumbo-sacral region. No blow anywhere. Much professed alarm and exaggeration about his injuries, and the day after the accident was sure he was going to have "paralysis of the spine of his back," and that he was dying and must make his will. Temperature and pulse never abnormal. After return home became more nervous than before. The only objective sign was an occasional slow pulse of 55. Change of air six months after accident did him no good, but he made no effort to do anything, got up late, went to bed early, stayed indoors all day, and nursed himself into a weakly state
137	M.	18	Same	Had a bruise on one shoulder. Later he complained of pain across the lower dorsal vertebræ, with tenderness there. He was much upset and was very prone to cry
138	F., married	28	Carriage ran off line	An excitable, hysterical woman. Felt generally shaken, and almost immediately had pain in the small of the back. This it was that troubled her afterwards, and she was very nervous. Poor sleep. Quick pulse. In fifth month of pregnancy

APPENDIX 341

Date of settlement.	When last heard of after accident.	Condition when last heard of.	Evidences of injury to spinal cord or membranes.	Remarks.
10 mos.	6 years	Family trouble since settlement. Still weak and nervous, but looks and is greatly better. Subject to palpitation and alternate sensations of heat and cold. Varies much from day to day. Legs very soon fatigued. No sign of central nervous disease	None	A genuine case of extreme nervous prostration and concussion of brain. Probably permanent damage to nervous stability, but certainly no meningo-myelitis. At the menopause.
9 mos.	6 years	At his work as usual, and enjoying good health	None	Delay owing to largeness of claim.
9 mos.	6 years	In excellent health, and at his usual work	None	Exorbitant claim. Much exaggeration from the very first.
4 mos.	—	Left the neighbourhood and cannot be traced	None	
3 mos.	2¼ years	Was very nervous until confinement was over. Then began to improve, and is now quite well	None	Previous hæmoptysis. Dulness at one apex

APPENDIX

General outline of case.

Blow on back of head. Momentarily unconscious. Said he was struck all down the back. No sign of constitutional disturbance. Exaggeration at very early date. Made a large claim, and immediately put his case in hands of solicitors having a reputation for success against railway companies. Later, he complained of pain in the back, impaired memory, loss of sight, &c. Not one objective sign

Stunned by blow on left occiput. Much after pain. Heaviness and oppression of head. In about three weeks, when he began to move about, he complained that he could not see so well, and on examination an opinion was expressed that he had "hyperæmia and cloudiness of both optic discs." More careful examination, however, revealed high myopic astigmatism,' to which the appearances in the fundus were really due. Also had double cataracts, which might have been perhaps accelerated, though not caused, by the accident. Much mental distress from fear of losing sight altogether

Felt sick, but did not vomit. Bruise on leg and arm. Later complained of pain in back, and held himself in stooping posture; also of weight and bewilderment in the head, bad sleep, &c. Never a single objective sign of injury

Much frightened. Blow on forehead and knees. In bed ten weeks with pain in head and in sacral region. Became very nervous. Had much pain and hyperæsthesia about back and forehead, and considerable trouble from pain and stiffness of ligamentum patellæ. Much better at end of thirteen months. Between accident and settlement had conceived and given birth to a child

Formerly addicted to alcohol, but at time of accident said to have reformed. Stunned by blow on face. In bed seven weeks with lumbo-sacral pain. Later had so-called bilious attacks, and it was suspected that she had resumed her old ways. This was acknowledged by her solicitor the moment the trial was over

APPENDIX 343

Date of settlement.	When last heard of after accident.	Condition when last heard of.	Evidences of injury to spinal cord or membranes.	Remarks.
17 mos.	6 years	No report obtainable. His doctor, whose evidence was not what he wished, dare not go near him	None	Bankrupt at time of accident. Litigation. Jury reduced claim to one fourth. Based largely on statements proved in court to be absolutely false.
10 mos.	6 years	In good health. Cataract steadily advanced in one eye	None	Concussion of brain.
13 mos.	18 mos.	Resumed work after settlement, and soon married. Cannot be traced further	None	Gross exaggeration. Litigation. Went to trial as a case of "concussion of the spine," with small hope of recovery.
13 mos.	—	Very soon afterwards left the neighbourhood, and cannot be traced	None	Genuine case. Ligitation, to which she herself was no party. A lawyer's case.
13 mos.	5½ years	Lost her husband and married again. In good health	None	Very large claim. Litigation as a hopeless case of "concussion of the spine."

No. of case.	Sex.	Age.	Nature of accident.	General outline of case.
144	M.	42	Slight collision	Blow on left side. Laid up a few weeks with headache, palpitation, and want of sleep. In four months much better and at work again. Some months after accident had attack of hæmorrhoids, prostatic abscess, and retention. A stout, plethoric man
145	M.	30	Same as 133	Thought he must have been stunned, having had a cut on his forehead, and a slight blow on his back, which was subsequently stiff. For a few weeks had pain in his head and was nervous when alone or unoccupied. A month after the accident took to walking with a limp and in a stooping posture
146	M.	42	Same	Asleep. Felt "hurt in the small of the back," and had a blow over the left lower ribs. Was laid up a fortnight with pain in the side and head, giddiness, and feeling shaky. Mending rapidly at end of a month
147	F., married	36	Same	Blow on back, chest, and side. Dazed. Vomited. Did not lie up as she felt worse in bed. Legs and arms felt very weak and shaky; bad sleep, "loss of memory," and very nervous. Some exaggeration
148	M.	50	Same as 67	Almost directly after had pain in the small of the back which he had in consequence well rubbed. Felt dazed and sick. Though feeling stiff he went about much as usual, not thinking much of his injuries. Three months after accident great thirst led to examination of urine which was found to contain large quantities of sugar. Careful inquiry elicited that two months before the accident he had been extremely thirsty; was, however, then summer. Amount of sugar much influenced by diet. A year after accident 798 grains in the gallon. Never any paralytic symptoms. A perfectly sober man, but addicted to taking continuously and daily at every meal large quantities of alcohol
149	M.	20	Same	"A lad of an anxious turn of mind," said his father. Tall, thin, and weak. Blow on face and on shins. In bed a week, and indoors for another fortnight, suffering from his legs, head, and back. In five weeks went away for change of air, took a tour in fact, and apparently did too much, for on his way home he nearly fainted. About three months and a half after

APPENDIX

Condition when last heard of.	Evidences of injury to spinal cord or membranes.	Remarks.
At business, but not capable of the same physical exertion. Still has some urinary trouble unconnected with injuries received	None	Previous ill-health. Moderate claim. No exaggeration.
Walked naturally directly his claim was settled. In perfect health since	None	Slight concussion. Much exaggeration.
No report obtainable	None	
No report obtainable	None	
After exposure to cold died of acute pneumonia. His doctor regarded his death as in no wise connected with the diabetes, for which he had not recently been under treatment. The man had been a very free liver	None	There was no evidence of nerve lesion in this case. Glycosuria probably existed before accident. A case wholly unlike that of traumatic diabetes published by Dr Buzzard, Clinical Society's 'Transactions,' vol. ix, p. 145, q. v.
Soon recovered after his claim was settled, and has been in continued good health since	None	A typical example of the origin of hysterical attacks in a neurotic male from syncope, the result of shock and fatigue.

APPENDIX

Sex.	Age.	Nature of accident.	General outline of case.
			the accident returned to work, but in a few days had another syncopal attack, and at once went off into hysterics. From that time forth had repeated hysterical seizures which were largely kept up and induced by the great indiscretion of his friends. Outrageous claim, and long delay in settlement
M.	47	Same as 120	Much and severely bruised and burned about face and body. Fracture of one fibula, and a compound fracture of both bones of the other leg just above the ankle. Much collapse. Rallied, however, and did well. Six months after the accident the note was: "He is well nourished; he has no nervous symptoms; his pulse is natural; sleep and appetite are good, and apart from the leg he is virtually well. He thinks that perhaps his memory is not quite so good." Adhesions about ankle very troublesome
M.	46	Same	Bruised below one knee. Complained of being slightly nervous, that his heart fluttered, and that he perspired too freely. Sleep good
M.	20	Same	Bruised on shin and shoulder. Felt rather stiff, and did not sleep well for two or three nights
M.	22	Same	Momentarily stunned by blow on side of head. Next morning vomited. Felt very shaky and nervous for a few days. After rest was much better
M.	35	Same	Asleep. Probably stunned. Large lacerated wound on forehead; suppurated. Felt very shaken and weak, but there was not much constitutional disturbance
M.	30	Same	Thrown down on floor of carriage. Cut and bruised about arm and hand. Went to work next day, but broke down, feeling so giddy. Became very nervous and restless, but after a few days' rest began to mend. Some years before had been laid up from anxiety and over-work
M.	30	Same	A singularly frail, feeble man. Blow on back of head almost stunned him. Blow on left side and on lower part of back. The excitement kept him up and he came home. Then he broke down and complained of excessive nervousness,

APPENDIX 347

Date of settlement.	When last heard of after accident.	Condition when last heard of.	Evidences of injury to spinal cord or membranes.	Remarks.
13 mos.	5 years	Still a little lame, and his legs pain him in cold weather. Otherwise, both mentally and bodily, he is in as good health as ever he was, and feels no ill-effects from the accident	None	Large claim. Large payment. No exaggeration. Broken legs. Much collapse. No after-symptoms of "nervous shock."
6 weeks	6 years	Medical report not possible. Is, however, at his usual work, and to all appearances well	None	Moderate claim. No exaggeration.
A week	—	A servant. Cannot be traced	None	
2 weeks	—	A servant. Cannot be traced	None	Slight concussion of brain.
6 weeks	2½ years	Constantly at work since. In perfect health	None	Small claim. No exaggeration.
6 weeks	2½ years	Feels rather nervous when travelling by rail, but is otherwise in perfect health	None	Moderate claim. In good circumstances.
3 mos.	—	Left the neighbourhood some time after, and cannot be traced	None	Previous ill-health. Very large claim. In pecuniary difficulties.

No. of case.	Sex.	Age.	Nature of accident.	General outline of case.
157	F., married	36	Same	a tendency to cry, and sensation of faintness and of being alternately hot and cold. Weak pulse, pupils wide. Much troubled for some weeks by morning sickness. Old attack of hæmorrhoids brought on. At end of two months decidedly better, and went away for change Pregnant. Stunned by blow on head. In bed a fortnight; suffered from pain in the head and much in the small of the back, especially on movement. Hysterical exaggeration. Not much constitutional disturbance. At end of four months seemed pretty well
158	M.	39	Same	Not stunned. Bruised about knees. Felt very nervous and shaky, and had pain in small of back. No constitutional disturbance. In a fortnight looked quite well
159	M.	31	Same as 132	Dazed. Blow on back and on back of head; also had one shoulder severely sprained. Absent from business for two months, when he partially resumed it. Five months afterwards still felt very nervous and confused, and was afraid to travel
160	M.	31	Sudden stoppage from break action at night	Said he was thrown off the seat, but had no bodily hurt. Must have been very frightened, for he "felt at once as if his heart was coming into his mouth." In bed two days. Sleepless, vomiting, and very nervous. Later complained of being easily startled and having pain in the lower part of the back
161	M.	46	Same as 120	Blow on legs and also on nape of neck. Very stiff at these parts. Became very nervous and shaky, sleepless and confused. Manner very agitated. In about six weeks was able to resume work
162	M.	45	Same	Blow on the back of the head, and for two or three days felt shaken and had much pain. In eleven days was perfectly well and at work again
163	M.	43	Same	Deeply incised wound, two inches long, over right parietal eminence. Dazed, but not unconscious. Considerable nervous shock afterwards; sleepless, weak pulse, pain down the back. Wound healed well. Two months after accident had much pain and tenderness at site of the scar, and complained of occasionally losing

Condition when last heard of.	Evidences of injury to spinal cord or membranes.	Remarks.
Confinement natural. No further medical attendance afterwards. Shortly after left neighbourhood and cannot be traced	None	
A labourer. Cannot be traced	None	
For some months after settlement nervous and afraid to drive. Now in good health	None	Moderate claim. Some unconscious exaggeration.
In good health	None	Uncomplicated fright.
Actively at work. Good health	None	Good circumstances. Small claim.
In good health, and at usual work	None	Good circumstances and position. Small claim.
At his usual work every day, and in good health. Has had no more fits	None	Jacksonian epilepsy.

No. of case.	Sex.	Age.	Nature of accident.	General outline of
				himself. Without going int said that all the symptoms poi ening of the meninges and pro cerebri at site of the blow. after accident was thought though his manner was som
164	M.	30	Same	Did not think he was hurt, bu head. Went to business next and was then laid up for a f in the head, sleeplessness, a ness, probably rheumatic, al
165	F., married	41	Same	Thought she was thrown to the although a stout, plethoric w on the top of the head; wa the left lower ribs, one of wh abdomen and the legs. ceased immediately. Confine and then became more ner wandering at night. This was kept up to a great exte and tenderness in region of abdominal fat prevented Never any elevation of temp returned in three months, tenderness was then lesseni
166	M.	29	Same	Dazed. No knowledge of blov a few days and felt shaken. and exposure at time of acc
167	F., married	25	Same	Two ribs broken close to their bruise over left sacro-iliac syr shock. Rallied, however, we Later became very nervous much hyperæsthesia at the p weeks after accident had an at night. Two months afte better. Pulse natural and disturbance. Complained over the 8th and 9th dorsal
168	M.	39	Same	No history of injury except a forehead and a severe bruis Subsequently some synoviti other symptoms

Date of settlement.	When last heard of after accident.	Condition when last heard of.	Evidences of injury to spinal cord or membranes.	Remarks.
3 mos.	2½ years	Been in perfect health, and felt no ill-effects	None	
6 mos.	—	Cannot be traced	None	
7 weeks	2½ years	In good health since	None	Good circumstances. Moderate claim.
11 mos.	—	Shortly after settlement left the neighbourhood, and cannot be traced	None	Reasonable claim, and at date of settlement was nearly well.
5 mos.	—	Lives abroad. Cannot be traced	None	

No. of case.	Sex.	Age.	Nature of accident.	General outline of case.
169	M.	51	Collision	Had a very severe bruise on one thigh and lesser bruises on ribs and one arm. Very stout, plethoric man. No symptoms of nervous shock, but much pulled down by injury to leg which was swollen and stiff for long. Sea voyage restored him
170	M.	38	Same	Healthy, but had had a "liver" in India. No knowledge of being hurt. In half an hour felt sick. Queer all day when at work, and in the evening, vomited. Had pain in the head. Awakened the same night by "excruciating agony" in his back. Was more or less in bed for ten days with paroxysmal pain in the back, with much hyperæsthesia, from 10th dorsal vertebra downwards. Also complained that his legs felt "numb in the bones," but there was never any anæsthesia. Sweated much. Constipation and flatulence gave rise to sensations, very alarming to him, in his abdomen. Became quite hypochondriacal. Very anxious about his business. Pulse easily quickened under excitement. Abnormal sensations all over, and very alive to them. Temperature normal throughout. When at length induced to move about he improved and the pains in his back diminished
171	F., single	24	Same	Bruised on forehead, not stunned. Vomited same night. Work next day, but obliged to give up. Hysterical and nervous for a few days and then began to improve. At work again in a fortnight
172	M.	39	Same	Dazed by blow on forehead. Retched, but did not vomit. Much headache for a few days. Nervous and shaky for a fortnight or so. Then resumed work
173	F., married	40	Same	Delicate woman suffering from menorrhagia. Stunned from blow on forehead. Suffered afterwards from occipital pain, sleeplessness, and general upset. Did not lie up, and very soon attended to her household duties
174	M.	30	Same	Conscious only of a blow on shin, but soon was in a tremble all over. Two days after accident had much pain in back, darting and shooting when he moved. Tenderness also became marked on one shoulder, and the arm felt as if "electrified." Reactionary fever for three days. In about fourteen days was much better, and he then rapidly improved and resumed his work

Date of settlement.	When last heard of after accident.	Condition when last heard of.	Evidences of injury to spinal cord or membranes.	Remarks.
3 mos.	5 years	Perfectly well. Quite got over the accident	None	
9 mos.	2 years	Perfectly well at this date. Left the neighbourhood later and cannot be traced	None	Very large claim. Genuine hypochondriasis.
A month	—	Cannot be traced	None	Self dependent. Slight concussion of brain.
2 mos.	6 years	"Felt the effects" for a few months. Perfectly well since	None	Good circumstances. Moderate claim.
2 mos.	6 years	Thinks she began to work too soon and felt nervous and shaky for at least 18 months. Now perfectly well, though nervous when in a train	None	Previous ill-health. Contre-coup. No exaggeration.
6 weeks	7 mos.	No further medical treatment. At work and well since settlement. Cannot be traced later	None	Moderate claim. No exaggeration.

No. of case.	Sex.	Age.	Nature of accident.	General outline of case.
175	M.	45	Slight shunting collision	Knocked down. Blow over right lower ribs. Doubtful fracture. Was in bed three weeks, and was then to all intents and purposes well. But he then began to exaggerate, stayed at home, and professed utter inability to do anything. No sign of ill-health
176	M.	33	—	See page 233
177	F., married	47	Same as 133	Thrown on floor, but picked herself up directly. Was conscious of all that happened, and thought herself unhurt. After she got home became excited and complained of headache, and had queer sensations in the head, back, and elsewhere, and was very nervous and apprehensive. Polyuria for a week. Catamenia, then due, were delayed a week, but were subsequently regular. Began to move about in a month, and then became more nervous than before; had tremors and quiverings, felt "numb" in her arms and legs, became breathless when there was any noise, lost her memory, could not calculate, and was more troubled with presbyopia; had a nervous twitch of the head, and complained much of her spine. Nervous complaints innumerable. Long journeys to the seaside did not upset her, and the change did her good. Menopause a year after accident
178	M.	48	Same	Small wound over left forehead. Not stunned, but seemed a good deal shaken. At end of three weeks was much better. From that time, however, began to get worse, *i.e.* he complained of being aged, of having severe pain in his back, especially when stooping, of loss of memory, of feeling good for nothing, of tinglings in his legs. Pulse and temperature never abnormal. His book knowledge was remarkable
179	M.	29	Collision in station	Severe blow on right sacro-iliac region. Very soon had a sensation of pins and needles all over his body, and in arms and legs. Home next day. In bed fourteen weeks, suffering from head and from "heavy pains" in the back. His legs also felt numb and heavy, although he could move them perfectly well. Four months afterwards able to go away for change, and on return resumed work. Head-

APPENDIX 355

Date of settlement.	When last heard of after accident.	Condition when last heard of.	Evidences of injury to spinal cord or membranes.	Remarks.
4 mos.	2½ years	At usual work and in perfect health. Occasionally gouty	None	Much later exaggeration after his recovery from immediate injury.
—	—	—	None	
22 mos.	6 years	In good health and at her usual work. Has lost the nervous manner, but she is much more easily fatigued than she used to be. Figures especially worry her	None	Exorbitant claim. Litigation. Much exaggeration and indiscreet sympathy of friends.
10 mos.	6 years	Very soon at his usual work and been in good health since	None	Pecuniary difficulties. Outrageous claim. Gross exaggeration. Litigation. Verdict less than one twentieth of the amount claimed.
13 mos.	5 years	Medical report impossible, but it is learned that "the injured leg is not the same size as the other, and he cannot do much walking or standing without fatigue. Otherwise, however, well"	See outline	No exaggeration. Moderate claim. A case in all respects like that recorded in the text, page 130. Injury to pelvic plexuses. ?

No. of case.	Sex.	Age.	Nature of accident.	General outline of case.
180	M.	38	Same as 120	ache, however, returned, and he felt weak in the back, and his back was so painful that he took to his bed for six weeks to undergo a course of blistering of the spine. Seven months after accident much better, but had pain, occasionally severe, in small of back and sacrum, made worse by walking. Right leg also felt numb and heavy. Carried it stiffly in walking and wore heel of boot unduly. No wasting. Temperature of two limbs equal. Cremasteric reflex normal. Could perform any desired movement, but with some stiffness and sluggishness. Could not feel pain in right leg quite so well as in left. Remembered events of accident and did not feel much hurt. Next day had pains all across the loins, and much pain in the head; also had to strain unduly in micturition. Not laid up in bed. Much continued pain across small of back and iliac crests, where there were marks of bruising. Became very nervous, stayed indoors, lost his appetite, and looked very ill. Much trouble from constipation
181	M.	49	Same as 133	Said he had a blow on the left cheek and supposed he was stunned. Remembered, however, all that happened. Stayed in bed a month, and was suffering from constant muddle and daziness, from pain in the back, from snatchings and twitchings, and from feeling all to pieces. For a few days his temperature was raised. Three months after the accident he made a large claim, and then began to exaggerate and lie. Complained of spinal tenderness, very variable under examination, and of loss of power in his left arm, which he held rigidly to his side, and of loss of motion and sensation in the left leg. No objective signs. He professed, when so desired, absolute inability to move his left leg. Sleeplessness, alarm, loss of memory, &c. When case went to trial he was able to hang about the court like other witnesses, although he had been wholly unable to go out of doors for months past. Had a wondrous book knowledge of symptoms
182	M.	32	Rather sharp collision	Feeble, delicate man. Thrown forward and struck by another man in the belly. Much pain in abdomen. Vomited. Violent palpitation. In bed six weeks. Temperature raised

APPENDIX 357

Date of settlement.	When last heard of after accident.	Condition when last heard of.	Evidences of injury to spinal cord or membranes.	Remarks.
6 mos.	3 years	Complained much of his back for some time, but is now at work and in good health	None	Rheumatics not long before the accident. Large claim. Unconscious exaggeration.
18 mos.	5 years	Had perfectly recovered in a fortnight, and had the impudence to attribute his recovery to a Turkish bath. Still in perfect health. His doctor calls him "that artful man"	None	Gross fraud founded on trivial injury. Went to trial as a grave case of spinal injury, which might end fatally.
12 mos.	4 years	Still has occasional palpitation, but is very much better than he was	None	Exorbitant claim. Litigation. Of some interest in the history is the fact that this man was said

358   APPENDIX

No. of case.	Sex.	Age.	Nature of accident.	General outline of case.
183	M.	56	Collision	for three days. Much abdominal pain, constipation, nervousness, and weakness. Severe family trouble prostrated him two months after accident, when he was beginning to mend. Later general nervousness, weakness, and abdominal tenderness. Chief trouble was palpitation. No organic disease of heart. At end of year much better Had a fracture of one clavicle and a blow, which did not stun him, on the head. Also bruised about the side. Went home in a week. At time of accident was under treatment for occipital neuralgia. This was increased by the accident, and he also had much "rheumatic" pain about the injured shoulder. For some months felt easily tired, and could not work so long. Absent from work two months
184	F., married	37	Same	Jumped from carriage, and had severe bruise on side of face. Some shock, followed by excitement the same evening. Was pretty well again in a fortnight
185	F., single	19	Same	Jumped from the train. Broke the clavicle and bruised cheek and shoulder. Next morning hysterical. Was well in three weeks, but then got into lawyer's hands and was advised to commence an action at once. Then began to complain of pain in the back and limbs, inability to get about or do anything. She was, however, delivered from the hands of the lawyer, and was soon quite well
186	M.	56	Same	Bruised about neck, cheek, and hip. Did not think he was hurt, but soon felt stiff and sore about the neck. Broke down on third day. Suffered from stiffness of neck, tenderness of cervical vertebræ, was prone to cry, and for long seemed utterly upset, the least excitement sending his pulse up to 150. Never any elevation of temperature. Stiffness of neck chiefly marked when being examined, for automatically it moved almost naturally. Much headache and vertigo. Lost flesh, and looked aged and ill. Seemed to have no ability to do anything. No change in condition for several months. Frequent micturition at night; no albumen. High tension pulse. Excessively nervous and apprehensive

Date of settlement.	When last heard of after accident.	Condition when last heard of.	Evidences of injury to spinal cord or membranes.	Remarks.
				to have the power of stopping his heart. At any rate, the cardiac innervation was probably abnormal before the accident.
3 mos.	4½ years	At work continuously since. Good health. No ill effects from accident	None	No exaggeration. Rheumatic.
6 mos.	—	Cannot be traced	None	Absence of husband delayed settlement.
3 mos.	2 years	At her usual work, and in good health	None	Note the origin of "litigation symptoms."
11 mos.	4 years	No work for 2 years. Then began gardening, and for last year or more been at his old employment, that of a cooper. Cannot, however, do so much as before. Sometimes giddy when stooping. Still some stiffness of neck, and after very heavy work his arms and legs feel numb and tired. Pulse 90, regular	None	A genuine case of nervous prostration.

No. of case.	Sex.	Age.	Nature of accident.	General outline of case.
187	F., single	39	Same	Could give no account of what happened, having lost her senses from the fright. Had one sleepless night, and next day felt dazed and stupefied. Did not abstain from work. Made no claim
188	F., married	62	Slight collision	Seemed slightly shaken, and was "hysterical." In an hour went on home, but was sick on the way. In bed a fortnight with pains about the body, and feeling "numb in the left side." In a very nervous, excitable state, and full of complaints. Later felt pains and tinglings, and coldness of her limbs, and tenderness, especially over the lumbar vertebræ. Large claim, and case drifted on with innumerable complaints of twitchings, numbness, spinal pains, giddiness, headache. Never any objective signs. Pulse and temperature throughout normal
189	F., married	38	—	See page 226
190	F., married	31	Slight collision	Excitable, hysterical woman. Had bruises on shoulder and both elbows. Had, in consequence, tingling in distribution of one ulnar nerve. Laid up a few days. Sleepless and nervous, and a month afterwards was still suffering from pain in her head and in the small of her back
191	M.	7	Same	Stunned by blow on occiput. Vomited. In bed nine days. Later had gradually subsiding symptoms of concussion of brain
192	F., married	63	Rather sharp collision.	Severe bruise over right lower ribs, where she suffered severe pain for some time. Doubtful localised traumatic pleurisy. Also much bruise on one leg. In bed a fortnight. Had no symptoms of nervous shock
193	M.	57	Collision	Bruise on one shin, and slight bruise on small of back. Felt sick, but did not vomit. Not laid up. No after-symptoms, but many complaints
194	M.	58	—	See page 115

Date of settlement.	When last heard of after accident.	Condition when last heard of.	Evidences of injury to spinal cord or membranes.	Remarks.
—	—	A servant. Cannot be traced	None	
9 mos.	5 years	Recovered pretty soon after claim was settled, and had good health since. Attending to her business as usual	None	Exorbitant claim. Symptoms due thereto. Litigation avoided.
—	—	—	None	
2 mos.	4 years	Had trifling ailments since, but is otherwise in good health	None	Small claim.
2 mos.	4 years	In good health	None	Concussion of brain. Early recovery.
5 weeks	—	A foreigner. Cannot be traced	None	
4 weeks	—	Left the neighbourhood. Cannot be traced	None	Some exaggeration.
—	—	—	None	

No. of case.	Sex.	Age.	Nature of accident.	General outline of case.
195	M.	47	Slightest possible bump in station	A free liver. Alcohol. Could give no account of accident, which was unknown to any one except himself. Seems to have had a "bilious attack" after he got home, but had no medical attendance until three weeks after the accident. Then made a large claim on the grounds that his spine had received injury. There were never any objective signs, but he complained much of pain and aching in the lumbar region. Before the accident he had often suffered from lumbago
196	F., single	27	Sharp collision in shunting	Small wound on one eyebrow, which healed at once. Subsequently nervous, easily fatigued, afraid to travel. Never any symptoms
197	F., married	42	Collision in shunting	Severe blow on right side of head and cheek. Not stunned. Almost immediately afterwards found herself stone deaf on the injured side. Was far away from home and anxious about her business, to which, contrary to advice, she persisted in returning in four days
198	M.	34	Very bad collision	Was strained about left side of chest and shoulder, but had no marks of bruising. No evidence of collapse or constitutional disturbance immediately after accident, but said he felt nervous. Pains in side, and later in small of back, kept him away from work for three months. Was then much better, but made a large claim, and concurrently with the worry about this there was a return of his former pains and much complaint of nervousness, lack of energy, &c. No symptoms
199	M.	30	Same	Abrasion on forehead and simple fracture of one fibula at lower end. Slight collapse. Recovered in the usual way and usual time after such an injury. Abstained, however, from work because of pain in his leg. Bodily health perfectly good
200	M.	46	Same	See page 167
201	M.	34	Same	Great collapse. Amputation of thigh in twelve hours. Lived four days

hen last heard of after ccident.	Condition when last heard of.	Evidences of injury to spinal cord or membranes.	Remarks.
—	No report obtainable. He threatened his doctor shortly after the trial	None	Litigation. Gross imposture. Jury gave less than one tenth of the amount asked. Went to trial as a case of "concussion of the spine."
—	Cannot be traced	None	Exorbitant claim for supposed damage to personal appearance. A "professional beauty."
5 years	Deafness much less. Has felt very nervous ever since, but is well and at her usual work	None	Although urged not to do so persisted in settling her claim at once. Genuine case of severe injury. Probable permanent deafness.
5 years	At his usual work, and in good health	None	"Litigation symptoms."
5 years	At work soon after settlement. Good health since and now	None	Delay in settlement due to business dispute. Some exaggeration.
—	—	None	.
—	—		

No. of case.	Sex.	Age.	Nature of accident.	General outline of case.
202	M.	57	Same	Severe scalp wound. Unconscious for a few moments. Severe bruise on hand. Rallied well from the collapse, and the next day was on full diet, having slept well. He complained, however, of pain and tenderness in the lower cervical region. In a month, when his wounds were healed, he caught a bad cold. Having recovered from this he declined to resume work, and complained more than ever of queer sensations all over. Never any sign of ill-health
203	F., married	40	Same	Severe laceration of one foot and leg. Much contusion also of arms and right gluteal region. Moved about too soon, and wounds when nearly healed broke out again. Also had a severe attack of phlegmonous erysipelas. Wounds healed under iodide of potassium. Never any symptoms of general nervous shock
204	M.	19	Same	Stunned by blow and severe wound on left temple. Much blanched from hæmorrhage. Also much bruised on legs, and on one knee which had only just recovered from synovitis from sprain. Three months after accident still nervous and easily fatigued. At end of twelve months had regained his colour and was in usual good health
205	M.	26	Same	Severe blow on right cheek and temple. Momentarily stunned. Rallied well, and in three days was eating and sleeping naturally. Pain about zygoma, and difficulty in mastication for some weeks. Two months after accident fainted after a long and fatiguing drive, and from that time became nervous and made many more complaints. Much trouble from hypermetropic asthenopia and lumbar stiffness
206	F., widow	59	Same	Never very strong. Bruises about hips, and very severe over left knee. Much effusion. Confined to bed more than two months. Suffered subsequently from much stiffness of the injured joint. Rheumatism. Much pain also in small of back. Much better at date of settlement

Date of settlement.	When last heard of after accident.	Condition when last heard of.	Evidences of injury to spinal cord or membranes.	Remarks.
8 mos.	4 years	At his usual work and in good health. Unsteady and addicted to drink	None	Made an exorbitant claim. Much exaggeration.
8 mos.	4 years	In good health since	None	
6 mos.	4 years	In good health	None	In good circumstances. Reasonable claim.
7 mos.	4 years	At his usual work and in good health. Still some asthenopia	None	
7 mos.	4¼ years	Has never felt quite so strong, and her memory not so good as it was. Occasional pain in legs. Otherwise good health and at usual work	None	

No. of case.	Sex.	Age.	Nature of accident.	General outline of case.
207	F., single	56	Collision	The only person injured. Lacerated wound over right eyebrow. Periostitis and necrosis, portions of bone coming away for nearly twelve months. Became very nervous and hysterical, and complained greatly of pain in small of back. Had much lifting to do in the course of her work. Hypochondriacal
208	M.	40	Same as 198	Momentarily stunned by blow on forehead. Rallied well, but in two days complained of pain and tenderness across sacral and gluteal regions, of pain and tenderness in both legs, of feeling ill all over, and as if he had been knocked into an old man. Gradually mended, and in six weeks was at work again partially
209	M.	21	Same	Small wound between eyebrows. Stunned. Next day his back began to pain, and he was four days in bed in consequence. Soon began to go about, but for some weeks had aching in his back. At work and well in two months
210	M.	24	Same	Unconscious for five minutes from blow on right forehead. Severe wound on one shin. In two days his back began to feel sore and painful right across the sacrum and iliac bones. Pain and tenderness over lower dorsal vertebræ. Four months after accident still had pain in head, made worse by noise or excitement, and pain in back when stooping. Tenderness about right sacro-iliac synchondrosis, and occasional pain in course of right sciatic. Nervous and easily startled, and his face flushed very readily
211	M.	34	Same	Wound over left eyebrow. Stunned. In a few days began to have pain across small of back when moving. Nervous and easily startled. Absent from work six weeks
212	M.	61	Collision in station	Stunned from blow on forehead. In bed a month, suffering from pain in head, and also much pain in back, so that movement was difficult. Then began to go about. Six months after the accident he was, however, complaining more than before of feeling utterly good for nothing, of being nervous when alone, and of much pain and stiffness of the back, with great hyperæsthesia. No objective signs. A free liver and frequenter of public houses
213	F., single	23	Very slight collision	Slightly bruised, and was much frightened. Had a bad night, and next day became very nervous and hysterical. Thus she continued for

APPENDIX

ate of ettle- ent.	When last heard of after accident.	Condition when last heard of.	Evidences of injury to spinal cord or membranes.	Remarks.
20 mos.	5 years	In good bodily health, but still complaining of her back. No symptoms of nerve disease	None	
5 mos.	3 years	Never looked quite so well, but had been continuously at work until 3 years after, when he died after Amussat's operation for obstruction caused by malignant disease	None	
4 mos.	4 years	At work since and in good health	None	No exaggeration.
6 mos.	4 years	Soon got well after settlement and has been at work and in good health since	None	Moderate claim. No exaggeration. Concussion of brain.
6 weeks	4 years	Recovered soon after settlement. In good health since	None	Small claim. No exaggeration.
8 mos.	—	No fixed abode. Cannot be traced	None	Exorbitant claim. Litigation. Settlement out of court for one third of amount asked. Much exaggeration. Pecuniary difficulties.
7 mos.	4¾ years	Within a very short time of settlement was going about as	None	Exaggerated claim. Physically and mentally the girl was unsuited for the

No. of case.	Sex.	Age.	Nature of accident.	General outline of case.
				several days making complaints which alarmed her friends. A leading examination then directed attention to her back, which became exceedingly hyperæsthetic. A volume might be filled with her complaints. Was delicate and puny. Never any functional derangement. Condition kept up by indiscreet sympathy of friends, who thought by compensation to extricate her from pecuniary difficulties
214	M.	49	Same as 198	A man of highly nervous temperament, though strong and powerful in build. Thrown on floor of carriage. Unconscious from fright, and had no recollection of the accident whatever. Blow over left lower ribs. Much pain there, and doubtful pleurisy for ten days. Pulse quick for a few days, but temperature not raised. Gradually improved, but two months after accident got into a most hysterical state, had sacral pain, stayed on couch all day, gasped and sighed when spoken to, and bitterly complained of his ailments. Had also a nervous cough, which incessantly troubled him. No later objective signs
215	F., single	30	Same as 196	Stunned by blow over left eyebrow. Continued her journey, contrary to advice, and was then laid up ten days in bed. Periostitis slight at site of blow. Two months afterwards still suffering from her head. Occupation brought on pain, and she was easily fatigued. Also very nervous.
216	M.	30	—	See p. 242.
217	F., single	33	Same as 196	Blow on one cheek and left side of back of neck. Much shaken and laid up for several days. Although feeling unfit she resumed her work in six weeks, but found herself unable to continue it to the full. Ten weeks after accident still much pain in cheek, and also in region of left great occipital nerve. Also heat and cold alternately of face and extremities. Much anxiety about her work.
218	M.	32	Same as 212	Seen instantly after the accident, he showed little sign of injury, but complained of his wrist and neck which he thought had been sprained. For a week or so complained of being nervous and dizzy, of pain in the head, and of being very despondent. After six

Date of settle- ment.	When last heard of after accident.	Condition when last heard of.	Evidences of injury to spinal cord or membranes.	Remarks.
		usual. In good health since		work in which she was engaged, and she was in pecuniary difficulties at time of accident
10 mos.	5 years	Looks and is in good health and at his usual work. Says it was getting on for three years before he was fit for work. Has pain in his side now when overworked	None	Made a most exorbitant claim. Much conscious and unconscious exaggeration
10 weeks	—	No settled home. Cannot be traced	None	Concussion of brain.
—	—	—	None	
4 mos.	5 years	In perfect health and at usual occupation as before the accident	None	No exaggeration.
8 mos.	19 mos.	"Recovered" directly his claim was settled. Been well since. Cannot be traced later than nineteen months	None	Exorbitant claim. Gross exaggeration. Pecuniary difficulties.

24

No. of case.	Sex.	Age.	Nature of accident.	General outline of case.
				weeks at a hydropathic establishment he resumed work. Made no complaint, and said it was a pleasure to be again occupied. Then learning that Case 212 had made a large claim, he thought he ought himself to make a still larger. From that time began to complain more than before of loss of memory, utter unfitness for work, of pain and throbbing in the head, and great despondency. Family history bad
219	M.	38	Same as 198	Momentarily stunned by blow on back of head, and had a graze on right malar eminence. Did not feel much amiss until the next evening, when he felt sore all over. Later he complained of pain, especially on lateral movement, in the cervical region and shoulders. No constitutional disturbance
220	M.	52	Same	Unconscious. Blow over left eyebrow, and a bruise of one knee. In two days felt worse and had pain all down the back, especially when he moved. A month afterwards looked pale and ill, complained much of sleeplessness, headache, nausea, constipation, coldness of his legs, and much hyperæsthesia of the back. Obviously severely shaken
221	F., married	47	Same	Blow between eyebrows. Felt shaken for a couple of days. Later complained of her back, of feeling nervous and easily startled. No constitutional disturbance. Well some time before settlement
222	M.	65	Same	Stunned by blow across forehead, and had a wound near one eyebrow. In bed a week. Much pain in back, right erector muscles chiefly, when he began to move about. Pain in head, giddiness, inability to apply himself, and much confusion of names and places, and a feeling of numbness down one side; weakness of voice for some weeks, constipation, frequent pulse, atheroma. At end of six months much better, and considering his age had done very well
223	M.	43	Rather sharp collision at night	Under treatment before accident for old stricture and occasional pus in urine. Blow on forehead but not stunned. Thrown on floor of carriage. Did not think himself much hurt. Was in bed

APPENDIX 371

Date of settlement.	When last heard of after accident	Condition when last heard of.	Evidences of injury to spinal cord or membranes.	Remarks.
4 mos.	4 years	Resumed work directly claim was settled. In good health since, but more excitable and more susceptible to influence of alcohol, in which he occasionally indulges	None	Considerable exaggeration.
9 mos.	5 years	Never been so well since or able to do so much work. Has some cardiac disease. No paralytic symptoms of any kind	None	A delicate nervous man before the accident, and constantly ailing.
9 mos.	4 years	Feels no ill effects from the accident	None	Slight concussion of brain.
7 mos.	4¾ years	Felt the effects for some time, but is now in excellent health, and attending daily to business. Failing a little from age during past 12 months. When much fatigued has pain in the back	None	No exaggeration.
14 mos.	4 years	Soon after settlement looked a totally different man and walked briskly and well. Con-	None	Exorbitant claim. Litigation. Settled out of court for one sixth of the original amount asked.

No. of case.	Sex.	Age.	Nature of accident.	General outline of case.
				a fortnight with pain in head and side. Pain in right lumbo-sacral region made him walk in a stooping posture for a long time. He became very nervous, easily startled, lost appetite and flesh, and had much hyperæsthesia about the dorsal spines, and looked languid, pale, and ill. Much worry from expected litigation. He attempted work about eight months after the accident. It was thought by some that he wilfully exaggerated
224	M.	56	Collision at night	Asleep. Much frightened and confused, but not hurt. At work next day, but in the evening felt "agitation in the nerves." Indoors four weeks complaining of bad sleep, loss of appetite, pain in the back, constipation, loss of memory, irritability of temper, inability to calculate, and general despondency. Also spoke much of creeping sensations under the skin. Never any sign of ill-health. Several months after the accident he affected to discover an old hydrocele and double rupture, caused, he said, by the accident. No evidence whatever to support the statement
225	F., single	22	Slight collision in station	Stunned by blow over right eyebrow. In bed ten days. Suffered much from pain in the head and giddiness, but more especially from hypermetropic asthenopia
226	F., single	34	Same	Wound over left eyebrow bled profusely. Bruise on left cheek. Laid up fourteen days. Subsequently suffered much from headache, giddiness, and palpitation. Very nervous. Much unconscious exaggeration
227	M.	30	Same as 198	Could give no account himself, but he had been terribly frightened, for he stood screaming near the train. Blow on right side of face, and for some months difficulty and pain in mastication. Also pain in head. Knocked up by long journey a month after accident, and became more nervous and sleepless. Much better at time of settlement
228	M.	45	Same	Conscious of a blow across the shoulders and back, and felt so stiff that he could give no help to others. Indoors a fortnight with pain and stiff-

APPENDIX. 373

Date of settlement.	When last heard of after accident.	Condition when last heard of.	Evidences of injury to spinal cord or membranes.	Remarks.
		tinuously at work and in good health since		
12 mos.	4 years	At work ever since, and in usual good health	None	Bankrupt. Exorbitant claim. Litigation. Settled out of court for less than one fourth of the amount claimed. Gross exaggeration.
7 mos.	5 years	Has had good health ever since, but is still troubled with asthenopia	None	In good circumstances. No exaggeration.
9 mos.	5 years	Is in excellent health	None	Slight concussion of brain and hæmorrhage.
6 mos.	5 years	In good health. Had no ailment since	None	
6 mos.	4 years	Feels absolutely no ill effects from the accident	None	Good circumstances. No exaggeration.

No. of case.	Sex.	Age.	Nature of accident.	General outline of case.
				ness in upper dorsal region, where there were marks of bruises. Then attempted work, which brought on pain in the head, singing in the ears, and general nervousness. In three months was hunting again, though more easily fatigued than formerly
229	M.	46	Overturned carriage	Blow on left side of forehead and insensible for five minutes. Vomited. Laid up for three weeks, suffering from pain in the head, feeling sore all over; had a catching pain in the small of the back which made stooping difficult, and general nervousness. From the first he wore a shade over the left eye in consequence of photophobia. Six months after the accident there was found much loss of acuity of vision in the left eye, limitation of the field, and slight changes in the tint of both discs, with some obscurity of their margins
230	F., married	27	Very slight collision	Feeble anæmic woman. Three years before had been in an accident and felt the effects of it for eight months. Struck on head by falling lamp. Not stunned. Laid up with headache, pain all down the back, singing in the ears, and great nervousness. Course of pregnancy not interfered with. Two months after accident was up and about, but then professed complete forgetfulness of everything in her past life, and even of her own name. Could give, however, a full account of all her sufferings. Very nervous, quick pulse, pale and ill. Labour natural six months after accident. Gradual improvement afterwards. Then large claim made. No attempt to resume household work. A year after accident was worse than before, having now become subject to hysterical fits, in which she would perform any suggested act or movement, although apparently unconscious. Frequently threatened, but never perpetrated, violence, and never hurt herself. After former accident had had precisely similar attacks. Much indiscreet and culpable sympathy of friends
231	F., married	30	Same	Mother in an asylum. Wound on forehead bled profusely. Went on well for five days, then fainted, and took to her bed. Much headache, giddiness, disturbed sleep, and general tremor under the least excitement; also uncontrollable

APPENDIX 375

Date of settlement.	When last heard of after accident.	Condition when last heard of.	Evidences of injury to spinal cord or membranes.	Remarks.
8 mos.	—	Cannot be traced	None	Case is recorded by Dr Gowers in his 'Medical Ophthalmoscopy,' edit. ii, p. 348, Case 59. "*Failure of sight with concentric limitation of fields after an injury to the head in a railway accident; congestion of discs,*" q. v.
18 mos.	4¾ years	In fair bodily health, and at work in shop. Had one child since settlement, and now again pregnant. Still occasional fits, account of which looks like true epilepsy. Complains of loss of memory, bad nights, and pain in the head. No mental defect. Husband's intemperance adds to her trouble	None	Exorbitant claim. Litigation. Jury gave one third of the amount asked, or the precise sum which the railway company had offered *twelve months before the trial*.
7 mos.	4¼ years	Attending to her work, and not again under medical care. Regained speech, but otherwise in much the same state.	None	Note the family history. Improvement was coincident with settlement of claim, which, however, had had no appa-

Age.	Nature of accident.	General outline of case.
		stammer. Two months after accident in most wretched state of prostration and tremor, and was hardly able to speak. This lasted three months. Had throughout a strange, vacant expression, and it was feared she might become insane. Without apparent reason she began to improve seven months after the accident
34	Same as 197	Blow on side of face and head, and momentarily stunned. Slept badly same night, and next day felt shaken. Went to work. Obliged, however, to give up, and was laid up fourteen days with pain in head and jaw and across the dorsal region. Any attempt at work worried him much, but exercise in the country did him good. Six months after accident still complaining of headache after work, of being easily tired, and of flushing of the face from trivial causes
14	Same as 198	Stunned by severe wound and blow over left eyebrow. In bed fourteen days having much pain in head. When he began to move about found he could not see so well as before with his left eye, although for a fortnight he had been able to see quite well. Right eye lost, but not removed, four years before from blow. Examination of left eye revealed distinct evidences of old choroido-retinitis. Without going into all the pathological changes discovered, the conclusion arrived at was that the earlier attacks had been very mild, and "being peripheral had caused no noticeable impairment of vision, and that the shock of the blow had aggravated the disease and favoured the occurrence of opacities in the vitreous." All parts of the equator were equally affected, which seemed to be against the likelihood of the changes having been solely produced by concussion.
38	Same	Momentarily stunned by blow on back of head. In bed ten days, and in another fortnight was able to resume work. He suffered, however, from pain and queer sensations in the head, and from pain all down the back, such as he had once experienced when lifting a heavy weight. Five months after accident still complained of headache and of being easily tired, but he was well nourished and looked in good health

Date of settlement.	When last heard of after accident.	Condition when last heard of.	Evidences of injury to spinal cord or membranes.	Remarks.
		Excitable, and complaining at times of her head. Declares she will end her days in an asylum		rent influence over her condition.
9 mos.	4 years	In good health, but from overwork may have headache and flushing of the face. This much less than it was. Suffered a good deal for some months after the trial	None	Concussion of brain. No exaggeration of symptoms. Large claim. Litigation.
12 mos.	4½ years	General health good. Eyesight unchanged	None	Still later evidence is to the effect that the condition really preceded the accident.
7 mos.	5 years	At his usual work, and in excellent health	None	Some exaggeration.

INDEX OF SUBJECTS

A

	PAGE
ABERCROMBIE on concussion of the spinal cord, with cases	9, 34, 47
Abortion, rarity of, from railway concussion . . .	182
Accidents, railway, alarming nature of . . .	162
—— importance of learning nature of . . .	269
Accommodation, weakened, in general nervous shock .	180, 208
Ages, effect of shock at different	169
Aggravated diseases	247
Alcohol in general nervous shock	203
—— susceptibility to, after brain concussion . .	187
ALLBUTT, Dr Clifford, optic changes in spinal diseases .	209
ALTHAUS, Dr, on the treatment of brain disease by the voltaic current	225
Anæmia, spinal	206
—— transient local, in general nervous shock . .	195
Anæsthesia, in general nervous shock . . .	194
—— of the skin, from disuse of limbs . . .	290
Analgesia, hysterical, in children	256
Analogy between concussion of the brain and concussion of the spinal cord . . .	7, 17, 23, 106, 110
Angular curvature, case of congenital . . .	140
Aphonia from severe nervous shock . . .	232
Appendix of cases	296, et seq.
Asthenopia in general nervous shock . .	180, 208
Asymmetry, bodily, in malingering . . .	254
Ataxy, locomotor (see Tabes dorsalis).	
Atrophy, optic, in nervous diseases . .	208, 209
Atropine, use of, for purposes of fraud . .	251, 252
Attention, effects of, on bodily sensations . . .	194
—— Paget upon	196

B

	PAGE
Back, cause of aching	115
—— hyperæsthesia of the	146
BASTIAN, Dr, case of concussion-lesion of spinal cord	52

BELL, Sir Charles, upon inflammation of the spinal marrow
    following injury to membranes . . 10, 42, 68
Belladonna, use of, for purposes of fraud . . 251, 252
Bladder, atony of, after shock 182
—— the "nervous" 121
Bones, fractures of, influence of, on general nervous shock 84, 85, 200
Bowel, hæmorrhage from the, in maligering . . 250
BOYER on concussion of the cord, with cases . . 12
Brachial plexus, Mr Erichsen's case of injury of the . . 81
Brain, analogy between concussion of, and concussion of cord
    7, 17, 23, 106, 110
  —— classification of concussion injuries of . . . 28
  —— coma and shock from concussion of . . 23, 24
  —— concussion of, Brodie upon injuries in . . 7
  —— fatigue of, in general nervous shock . . 176, 177
  —— frequency of lesions in severe concussion of . . 27
  —— Hewett upon lesions in concussion of . . . 24
  —— Hutchinson upon lesions in concussion of . . 27
  —— physical surroundings of the . . . 30
  —— præfrontal lesions of, Ferrier upon . . . 26
  —— table of symptoms of concussion of . . 186, 187
BRIQUET on predisposition to hysteria . . . 215
BRODIE, Sir Benjamin, on injuries to the spinal cord . . 3
  —— on the analogy between concussion of the brain and spinal
    cord 7
Bromide of potassium, misuse of . . . 202
Burnings in the gouty 204
—— sensations of, in nervous shock . . . 194
Bursæ, distended, in malingering . . . 253
BUZZARD, Dr, a case of hysterical hemianæsthesia . . 224
—— a case of paralysis of one leg . . . 132
—— on cases of syphilitic paraplegia . . . 137
—— on recurrent herpes in tabes . . . 128
—— on the use of faradaism . . . 289

C

	PAGE
Caries, spinal, rarity of, from railway injury	139
Catalepsy, case of hypnotic	238
—— Heidenhain on	242
Catamenia, derangements of the, from shock	179, 182
Cauda equina, its physical surroundings	32, 49
Changes molecular, significance of the term	92
CHARCOT, M., on accidental discovery of hemianæsthesia	227
Chronic invalidism	199
Circulation, disturbances of, from shock	23, 174
—— state of the, in functional nervous disorders	228
CLARK, Mr Le Gros, on concussion of the spine	109
—— on general nervous shock	110
CLARKE, Mr Bruce-, on transient optic changes in spinal cord injuries	210
CLARKE, Lockhart, pathological report of concussion lesion of cord	101
CLINTON, Chief Justice, on expert medical evidence	294
Cold, heat and, sensations of, after shock	175, 195
Collapse (*see* Shock).	
Compensation	197, 199, 258, 279, 286
—— influence of, on the neuromimeses	222, 235
Concussion of the brain (*see* also Brain).	
—— Mr Hutchinson's table of the symptoms of	186, 187
"Concussion of the spine" (*see* also Spine).	
—— from slight or indirect injury, Mr Erichsen upon	83, 84
—— in association with hysteria	155
—— in malingering	266
—— leading to chronic meningitis and subacute myelitis, Mr Erichsen upon	86, 88, 93
—— Mr Erichsen's works upon	58
—— nature and extent of, Mr Erichsen upon the	67, 69
—— objections to the term	59, 82
—— the pathology of, Mr Erichsen upon the	91, 92, 98
Concussion of the spinal cord (*see* Cord).	
Constipation in lumbago	121
Constitutions predisposing to nervous disorders	204, 212, 216
Convalescence, delay in, from litigation	197
—— influence of compensation on	282
—— stage of, in general nervous shock	189
Convulsions, hypnotic	220

INDEX OF SUBJECTS

	PAGE
Convulsions, voluntarily induced	221
COOPER, Sir A., cases of concussion of the spinal marrow	70
Cord, the spinal, Abercrombie on concussion of, with cases	9, 34, 47
—— analogy of concussion of, with concussion of brain	7, 17, 23, 106, 110
—— Bastian's case of concussion of	52
—— Bell upon injury of	10, 42, 68
—— Boyer's cases of concussion of	12
—— Brodie's cases of concussion of	5
—— Brodie's classification of injuries of	3
—— concussion of, in military practice	20, 21
—— the spinal, Cooper's cases of concussion of	70
—— Curling's case of concussion of	46
—— Edmunds' case of concussion of	22
—— Erb on concussion of	111
—— gravity of injuries to	1, 42
—— Gull's, Sir W., cases of injury to	40, *et seq.*
—— hæmorrhage in substance of	39, 107
—— hæmorrhage upon the	14, 16, 38, 43, 77, 110
—— Hilton on concussion of	126
—— Hunter's case of concussion of	47
—— Hutchinson's case of recovery after concussion-lesion of	48
—— inflammation of the, after injury	10, 42, 48, 68, 133
—— laceration of, priapism from	80
—— liability to injury from injury to vertebral column, Mr Erichsen upon	60
—— Lidell on concussion of	15, 16, 35
—— Lochner's case of concussion of	46
—— Mayo upon concussion of, with cases	10
—— physical surroundings of	30, 31
—— popular belief in injury to, from railway collisions	2, 260
—— rarity of lesions of, from simple blows	32
—— rarity of recovery after undoubted lesions of	42, 48
—— remote effects of concussion upon, Mr Erichsen upon	69, 70, 84
—— Savory on concussion of the	35
—— severe bend of	39, 107
—— Sharkey's case of concussion of	50
—— Shaw on concussion of	106
—— stunning of, Dr Wilks upon	33, 148
—— Syme on concussion of	14
—— thickening of, from injury	66, 69

INDEX OF SUBJECTS 383

	PAGE
Cross-examination	. 291
CURLING, Mr, case of concussion of spinal cord	46
Cysts, sebaceous, in malingering	. 253

D

Danger of injuries to spinal cord	1, 16, 42, 48
Death from uncomplicated shock	184, 185
Deception, self-, in hysteria	. 256
Diarrhœa, from vaso-motor paresis	. 179
DONDERS, Prof., on muscæ volitantes	. 180
DONKIN, Dr, on tabes dorsalis	. 154
Dynamometer, use of the	. 288

E

Electricity, in a case of hemianæsthesia	. 224
—— in hysteria	. 289
—— in treatment of brain-lesions	. 225
"Electric test," the	271, 288
Emotional feelings, giving way to	. 205
Emotions, lost control of the	188, 191
Employers' Liability Act, note on the	. 258
EDMUNDS, Dr, case of concussion of cord	. 22
ERB, Prof., on concussion of the spinal cord	. 111
—— on spinal anæmia	. 207
ERICHSEN, Mr, on "concussion of the spine"	58, et passim
—— on meningo-myelitis mistaken for hysteria	. 217
Evangelicalism	. 277
Evidence, medical, in courts of law	. 292
Exaggerated diseases	. 247
Exaggeration in hysteria	196, 256, 275
Examination, methods of clinical	291, 292
Experiments, made by disease	. 106
—— objections to hypnotic	. 222
Eyesight, defects of, in general nervous shock	180, 208
—— in malingering	. 252

F

	PAGE
Facial nerve, peripheral paralysis of the	. 253
Factitious diseases	. 247
Family history predisposing to nerve disorder	214, 215
Fatigue, brain, in general nervous shock	176, 177
Feigned diseases, Ogston on	. 247
FERRIER, Dr, on præfrontal brain lesions	. 26
Fictitious diseases	. 247
FLETCHER, Dr, on a case of railway injury	. 337
Flushings in general nervous shock	176, 195
Fox, Dr Long, on the curability of tabes dorsalis	. 96
Fractures, absence of symptoms of "general nervous shock" after severe	84, 200
Fright, death from shock after	. 184
—— deferred syncope from	. 164
—— in inducing hysterical seizures	. 237
—— the cause of shock in collisions	162, 170
Functional nervous disorders (*see* also Neuromimeses)	
—— absence of organic disease in the	. 225
—— changes underlying	217, 218
—— convulsions purposely induced	. 221
—— predisposition to suffer from	. 212
Fundus oculi, changes in, in cord diseases	. 208

G

Galvanism (*see* Electricity).	
GAVIN, on tympanitis	. 250
Giddiness from disturbances of circulation	. 176
Globus hystericus	. 177
Gonorrhœa preceding spinal myelitis	. 135
GORE, on a case of railway injury	. 101
Gout, as a cause of cardiac disturbance	. 175
—— brought out by injury	. 104
—— in combination with nervous temperament	. 204
GOWERS, Dr, on degeneration of the posterior columns	. 153
—— on optic changes in spinal cord diseases	. 209
—— on pachymeningitis	. 137

	PAGE
GOWERS, Dr, on spinal anæmia	206
—— on state of the brain in functional paraplegia	230
—— on symptomatic value of spinal pain	141
GULL, Sir Wm., cases of paraplegia	40, *et seq.*
Gummata in imposture	253

H

Habit, morbid, of nervous system	205, 223
HAMMOND, Dr, on spinal anæmia	206
Hæmorrhage, as suggested cause of cord lesion in case of inherited syphilis	49
—— from the bowel in malingering	251
—— in centre of spinal cord	39, 107
—— in spinal canal from blows on spine	43, 71, 77
—— in spinal canal, Le Gros Clark upon	110
—— in supposed concussion of the spinal cord	13, 14, 38
—— intraspinal, preceding myelitis	135
Head, oppression and pain of, in nervous shock	176
Heart, palpitation of the	175
—— paresis of the, in shock	158, 159
Heat and cold, sensations of, in vaso-motor paresis	176, 195
HEIDENHAIN, Dr, on catalepsy	242
—— on hypnotism	222
Hemianæsthesia, Dr. Buzzard's case of	224
—— Dr Putnam on	274
Hernia in imposture	253
Herpes zoster	128
HEWETT, Sir Prescott, on lesions of brain from concussion	24
HILTON, on concussion of the spinal cord	126
—— on the position of inflamed joints	248
HODGES, Dr R. M., on sprains of the vertebral ligaments	156
HOOD, Mr Wharton, on traumatic lumbago	121
Hospital patients, free from compensation for injuries	280
—— lessons to be learned from	258, 279
HUNTER, case of spinal myelitis from injury	47
HUTCHINSON, Mr Jonathan, a case of recovery after concussion-lesion of the marrow	48
—— cases of brain concussion	27
—— on a case of dislocation of the spine	38
—— on the size of the pupil	181

INDEX OF SUBJECTS

	PAGE
HUTCHINSON, Mr Jonathan, on the symptoms of brain-concussion	186, 187
—— on the use of atropine	251
—— on xanthelasma	226
Hydrocele in imposture	253
Hyperæmia, transient local, in general nervous shock	195
—— spinal	206
Hyperæsthesia, a symptom of "concussion of the spine," Mr Erichsen upon	78
—— functional, Dr Wilks upon	142
—— in general nervous shock	194
—— in pachymeningitis, Dr Gowers upon	137
—— of back after railway injury	146
Hyperexcitability and paresis underlying functional disorders	218
Hypermetropia revealed by nervous shock	180
Hypnotism, objection to experiments in	222
—— state of the sensorium in	219, 242
—— voluntarily induced	221
Hypochondriasis	193
Hysteria, abeyance of will in	207, 256
—— anæsthesia of skin in	274, 290
—— exaggeration in	196, 256, 275
—— in connection with spinal injuries	155
—— in men	188
—— mistaken for myelitis	217, 244
—— nature of	191
—— paraplegia in	229
Hysteria, predisposition to	212, 215
—— seizures in	221, 235
—— use of electricity in	289

I

Imitation of nerve diseases (*see* also Mimicry)	212
Inflammation, of spinal membranes (*see* also Membranes)	68, 84, 86, 135, 139, 153
—— of spinal cord	10, 42, 48, 68, 133
"Inflammatory irritation of the membranes"	245
Inheritance in nervous mimicry	214
Injury revealing constitutional disease	105

	PAGE
Insanity predisposing to mimicry	215, 221, 224, 241, 244
Invalidism, causes of chronic	199

J

Jar, analogy of falling watch with, Mr Erichsen's	85
—— not felt in cases of fracture	84
—— of every organ	156, 270
—— sleep a protection against effects of	85
—— vibratory, in railway accidents, Mr Erichsen upon	70, 85
JOHNSON, Dr George, on backache	115
Joints, case of hysterical disease of	257
—— sprains of, comparison with spinal sprains	122, 144
—— stiffness of, in malingering	249, 267
—— the position of inflamed	248
—— treatment of sprains of spinal	144
JONES, Dr Handfield, on hyperexcitability and paresis	218
—— on secretion of sweat	178
JONES, Mr Wharton, on failure of sight from railway injuries, &c.	211
JORDAN, Mr Furneaux, on death from uncomplicated shock	185
—— on deferred shock	164
—— on hysteria in men	188
—— on shock at various ages	169
—— on shock from fright	163
—— on the circulation in shock	160

K

Knee-joint, simulated disease of	248, 267

L

Lactation, excessive, from vaso-motor disturbance	179
LAYCOCK, on pain and tenderness of the spine	119
—— on the neuromimetic convulsions	220
Leading questions, use of, in examination of patients	287
LIDELL, Dr, on concussion of the spinal cord with cases	15, 16, 35
Ligaments, Hodges on sprains of spinal	156
—— sprains of spinal	114, 122, 126, 132
Limbs, position of, in joint disease	248
—— in malingering	267

	PAGE
Litigation, beneficial effects of close of	157
—— evil effects of, on general nervous shock	198
—— recovery after	97, 268, 273
"Litigation symptoms"	116, 198
LIVEING, Dr E., on emotional feelings	205
LOCHNER, Dr, case of concussion-lesion of cord	46
Locomotor ataxy (*see* Tabes dorsalis)	
LORDAT, Count de, case of the	64
Lumbago, traumatic	114, 118, 121
—— pseudo-palsy from	71, 120, 121
LUSH, Dr W. J. H., case of infantile paralysis	107

M

Malaise, induced	263, 266
Malingering, bodily asymmetry in	254
—— hæmorrhage in	251
—— precedent disease in	253
—— starvation in	259, 262, 267
—— stiffness of joints in	249, 267
—— sweating in	269
—— the difficulty of successfully	252
—— the morality of	275
—— the motive for, after railway accidents	258
—— the temptation to, after railway injury	246
—— tympanitis in	250
—— use of atropine in	251
MANSELL-MOULLIN, Mr C. W., on shock	160
MAUDSLEY, Dr, on evangelicalism	277
—— on healthy occupation	201
—— on hypochondriasis	194
—— on the organic or systemic impressions	192
—— on the nervous temperament	214
MAYO, H., case of intra-spinal hæmorrhage	43
—— on concussion of the spinal cord	10
Membranes spinal (*see* also Inflammation).	
—— cases of death from chronic inflammation of Mr Erichsen's	95
—— chronic inflammation of, Mr Erichsen upon	86, 93
—— inflammation of, from vibratory jar, Mr Erichsen upon	84
—— inflammation of preceding tabes dorsalis, Dr Gowers upon	153

INDEX OF SUBJECTS 389

	PAGE
Membranes, " inflammatory irritation " of the	. 245
—— symptoms of chronic inflammation of, Mr Erichsen upon the.	89
—— syphilis as a cause of inflammation of . .	69, 137
—— traumatic inflammation of . . .	10, 69, 135
—— prognosis after inflammation of . .	. 138
Memory, loss of	. 181
Meningitis, spinal (*see* also Membranes).	
—— pachy-, Dr Gowers upon . . .	. 137
—— rarity of, after railway injury . .	. 139
—— subacute, as supposed cause of symptoms after railway shock	
	136, 245
Menorrhagia	179, 182
Mesmerism, influence of, on the sensorium . .	. 219
Micturition, difficult, in lumbago . . .	71, 120
Mimicry, induced by shock . . .	. 215
—— in malingering	. 272
—— nervous	. 212
—— temperament predisposing to . .	204, 212, 241, 244
Mind, effect of bodily injury upon the . .	. 191
—— effect of the, upon pain . . .	. 196
Molecular changes, Mr Erichsen upon . .	. 92
—— not necessarily gross pathological . .	. 148
"Mors, nec silet," motto of Pathological Society .	. 91
Movement, fear of, in sprain of spine .	. 120
—— treatment by, of spinal sprains .	. 144
Moxon, Dr, on the nutrition of the spinal cord .	. 228
Muscæ volitantes	. 180
Muscles, nutrition of, in hysteria . . .	. 289
—— of spine, sprains of	114, 125
—— rupture of spinal, in sprain . . .	. 115
Mydriasis	. 251
Myelitis (*see* also Cord).	
—— fatal cases of spinal . .	. 6, 40, 45, 133
—— from indirect " concussion of the spine " .	. 90
—— inherited syphilis in case of spinal . .	. 48
—— mistaken for hysteria . . .	. 217
—— railway case of death from, Mr Erichsen's .	74, 95
—— subacute, Mr Erichsen upon . .	86, *et seq.*
—— the pathology of, Mr Erichsen upon . .	93, 94
—— the symptoms of subacute, Mr Erichsen upon .	. 89

N

	PAGE
Nerves, injury to trunks of	123, 126
—— injury to, of brachial plexus	. 81
—— injury to, of pelvis and thigh	. 130
—— involved in spinal sprains	. 126
Nervousness	. 177
Nervous shock (*see* also Shock).	
—— absence of symptoms of, after fracture	84, 200
—— general	. 158
—— in association with spinal sprains	116, 190, 238
—— leading to mimicry	. 215
—— Le Gros Clark upon universal	. 110
—— spurious	167, 266
——symptoms of general	173, *et seq.*
Neurasthenia, bromide of potassium in	. 202
Neuromimesis (*see* also Mimicry).	
—— association with spinal sprain	234, 240, 244
—— changes underlying	217, *et seq.*
—— convulsions in	. 221
—— hyperexcitability and paresis in	. 219
—— illustrative cases of	226, 230, 233, 235, 238, 242
—— induced by nervous shock	. 215
—— in malingering	. 272
—— physiological disturbances in	. 216
—— predisposition to exhibition of	212, 214
Numbness	. 123
—— value of term	. 127

O

Occupation, want of, a cause of ill health	199, 201, 244
Ogston, Prof., on feigned diseases	. 247
Old, influence of shock on the	. 169
Optic disc, transient changes in, in spinal cord diseases	. 210
Orchitis in a case of myelitis	. 135
Organic sensations, in general nervous shock	. 192
—— Sully and Maudsley upon the	192, 194
Ovarian derangements from shock	179, 182

P

	PAGE
PAGET, Sir James, on absence of organic changes in the neuro-mimeses	225
—— on combination of gouty and nervous constitutions	204
—— on inheritance of nervous mimicry	214
—— on mimicry	213
—— on vaso-motor disturbances	195
—— on weakness of the will	230
—— upon attention	196
Pain, a subordinate symptom in spinal disease	143
—— influenced by attention	196
—— in lumbago	118
—— nature of, in vertebral sprains	143
—— of the spine, Gowers and Wilks on	141
—— spinal, in association with nervous shock	116, 238
—— value of, in disease	248
—— value of spinal, as a symptom	120, 142
Palpitation	175
Panting	177
Paralysis, feignings of	252
—— peripheral facial	253
Paraplegia, case of functional	226
—— case of functional motor	230
—— from intra-spinal hæmorrhage	5, 13, 16, 38, 43, 77, 110
—— nutrition of cord in functional	228
—— syphilitic	138
Paresis, of the circulation	24, 158, 174
—— hyperexcitability and	219
Pathology of "concussion of the spine"	90, et seq.
Pelvis, injury to nerves of, &c.	131, 132
PETIT, Dr L. H., on traumatic origin of tabes dorsalis	151
Physical surroundings, of the brain	30
—— of the spinal cord	30, 31
"Pins and needles," abnormal sensations of	123, 125
—— Hilton upon	126
Plexus, brachial, case of injury to Mr Erichsen's	81
—— lumbar, injury to	132
Pneumogastric, nucleus of the, derangements of	179
Polyuria	178
Poro-plastic jackets, misuse of, in spinal sprains	144

	PAGE
Potassium bromide, misuse of	202
Pregnancy, effect of railway concussion upon	182
Priapism	80
Prisoners, malingering	247
—— influences of sentences on health of	286
Prognosis, after general nervous shock	203
—— after railway injuries of spine	148
—— of spinal meningitis	138
Pseudo-palsy from traumatic lumbago	71, 120, 156
Pulse, state of the, in general nervous shock	159, 175
Pupil, size of, in nervous shock	181
—— in malingering	251
PUTNAM, Dr, on hemianæsthesia	274

Q

Questions, leading, use of	287

R

Railway collision, its effect on the spine	114, 117, 125
—— collisions, alarming nature of	162
—— fright, the cause of shock in	162, 164, 170
—— injuries, recovery after	281
—— motive for malingering after	246, 258
"Railway spine," the	63, 122
Reaction after shock	161, 186
Recovery, after general nervous shock	203
—— after railway injuries	281
—— delay in, causes of	198, 201, 266
Reflection, on bodily sensations, influence of	196
Refraction, errors in, revealed by nervous shock	180, 208
Religious zeal, ill results of misguided	221
Retention of urine from shock	182
Rest in the treatment of concussion of the spinal cord, Hilton upon	126
Rigidity of limbs in malingering	249, 267
RIGLER, Dr, on the frequency of railway injury of the back	113
REYNOLDS, Dr Russell, on medical evidence in railway cases	293
Ross, Dr, on syphilis of the nervous system	138

S

	PAGE
SAVORY, Mr, on concussion of the cord	35
Secondary degeneration of cord from concussion lesion, Bastian's case	52
—— of cord from injury to membranes	66
—— of cord from supposed molecular disturbance	148
—— of cord, Gore's case of	101
—— of the posterior columns, Gower's upon	153
—— Shaw on Gore's case of	103
Sensations, abnormal, in spinal injuries	123
—— abnormal peripheral, in pachymeningitis	139
—— effect of attention on bodily	196
—— modifications of, from "concussion of the spine"	78
—— subjective in general nervous shock	173
—— subjective, in malingering	266
—— the organic	192, 194
Sensorium, state of the, in hysteria	216, 219, 230
SHARKEY, Dr, case of concussion of cord	50
SHAW, Mr, on "concussion of the spine"	59, 106
—— on Gore's case	103
—— on severe bend of the spine and spinal cord	39, 107
—— on spinal pain	143
Shock (*see also* Nervous shock).	
—— asthenopia after	180, 208
—— catamenial derangements after	179, 182
—— death from uncomplicated	184
—— deferred	164
—— disturbances of circulation after general nervous	174
—— effects of on heart and circulation	23, 159, 174
—— effects of railway, on pregnancy	182
—— evil effects of litigation on after-symptoms of	198
—— fracture and	84, 200
—— from fright	162, 170
—— Furneaux Jordan on	160, 163, 169
—— general nervous	158, *et seq.*
—— headache after	176
—— hysterical seizures after	221, 235
—— in association with spinal sprains	116, 190, 238
—— Le Gros Clark upon	110
—— loss of memory after	181

INDEX OF SUBJECTS

	PAGE
Shock, Mansell-Moullin upon	160
—— nervousness after	177
—— on predisposition to suffer from, at different ages	169
—— organic sensations in general nervous	192
—— polyuria, diarrhœa, and menorrhagia after	178
—— predisposing to nervous mimicry	215
—— retention of urine after	182
—— severity of, in railway accidents	162
—— sleeplessness after	173
—— spinal anæmia after	206
—— "spurious nervous"	167, 266
—— subjectivity of later symptoms of	173
—— sweating after	178
—— symptoms of, shortly after collisions	165, *et seq.*
—— synonymous with collapse	158
Sighing	177
Skin, anæsthesia of, from disuse of limbs	290
Sleep, a protection against "vibratory jar"	85
—— value of, as a symptom	174
Sleeplessness in general nervous shock	173, 199
Sodium, bromide of	203
Spasm, muscular, as a symptom of "concussion of the spine"	78
—— from pressure on the cord, Brodie upon	6
—— functional, of arm, case of	233
SPENCE on concussion of the brain	29
Sphincters, paralysis of	78
Spinal anæmia	206
Spinal cord (*see* Cord).	
Spine, angular curvature of, from severe bend	78
—— bend of, injury to cord from	39, 107
—— caries of, from sprains, rarity of	139
—— cervical, extreme bend of	39
—— "concussion of," from direct violence	77
—— "concussion of," from slight or indirect injury, Mr Erichsen upon	83, 84
—— "concussion of," Le Gros Clark upon	109
—— "concussion of," Mr Erichsen's typical case of	66
—— "concussion of," Mr Erichsen's works on	58
—— "concussion of," pathology of, Mr Erichsen upon	93, *et seq.*
—— "concussion of," symptoms of, Mr Erichsen upon	78
—— congenital deformity of, case of	140

INDEX OF SUBJECTS

	PAGE
Spine, death from "concussion of," Mr Erichsen's case	. 75
—— dislocation or fracture of, without deformity	. . 9, 38
—— effects of direct blows upon, Mr Erichsen's cases of	72, *et seq.*
—— effects of severe bend of . . .	39, 77, 107
—— frequency of injury to, in railway collisions	. . 113
—— Hood, Wharton, on sprains of . .	. 121
—— hyperæsthesia of	. 146
—— injury of, preceding tabes dorsalis . .	. 149
—— Laycock on examination of the . .	. 119
—— pain in severe sprains of . . .	118, 156
—— paralysis of one arm from "concussion of," Mr Erichsen's case	. 81
—— paraplegia from blows upon . .	5, 13, 43, 71, 77
—— possible injury to, in Gore's case . .	. 105
—— prognosis after injuries of . . .	148, 149
—— pseudo-palsy from sprains of . .	. 71, 120, 156
—— Shaw on "concussion" of the . .	59, 106
—— simple sprain of	. 114
—— sprain of, and nervous shock . .	. 116, 190, 238
—— stiffness of, after severe sprains . .	144, 156
—— tenderness in severe sprains of . .	118, 156
—— the "railway"	63, 122
—— treatment of severe sprains of . .	. 144
Sprains (*see* also Spine).	
—— of vertebral column . .	. 114, 117, 144
—— pseudo-palsy after severe spinal .	. 71, 120, 156
—— treatment of spinal . . .	. 144
Stammering	. 177
Starving in malingering . .	. 259, 262, 267
Stunning of the cord from blow, Dr Wilks upon	. 33, 148
Subjective sensations, after railway injuries	. 261, 266
—— in general nervous shock . . .	. 173
—— in malingering	. 266
SULLY, Mr, on the organic sensations .	. 192, 194
Sweating in general nervous shock . .	. 178
—— purposely induced	. 269
SYME on concussion of the spinal cord .	. 14
Syphilis, as a cause of spinal meningitis . .	69, 138
—— Dr Buzzard on, in a case of paraplegia .	. 137
—— inherited, in a case of concussion-lesion of spinal cord	. 48

T

	PAGE
Tabes dorsalis, first revealed by injury or shock	105
—— Donkin on the origin of	154
—— Gowers on antecedent changes in cord or membranes	153
—— injury as a cause of	149
—— Petit on the traumatic origin of	151
—— the curability of, Fox on	96
Temperament, the nervous	204, 213, 214
Temperature, modifications of, in " concussion of the spine "	78
Tenderness, in " traumatic lumbago "	118, 156
—— value of spinal, as a symptom	119, 122
" Test, the electric "	271, 288
Tetanus, simulated	287
Tingling, in the gouty	204
—— sensation of, after spinal injury	123
—— value of	127
" Traumatic lumbago "	114, 118, 121
Tuke, Dr Hack, on automatic cerebral action	207, 208
—— on fear	164
—— on influence of reflection on the bodily sensations	196
Tumours, fatty, in malingering	253
Tympanitis in malingering	250

U

Urine, retention of, from shock	182
Uterus, disturbance of pregnant, in concussion	182

V

Varicocele in imposture	253
Vaso-motor system, derangements of, from shock	176, 186, 195
—— polyuria and diarrhœa from paresis of	178
—— relation of the, to the emotions	179
—— sweating from paresis of	177
Vertebral column (*see* Spine).	
Vision, defects of, after shock	180, 208

	PAGE
Vision, in malingering	. 252
Voice, feebleness of, after shock	. 177
Vomiting, after shock	166, 186
—— in hysteria and malingering	. 268

W

WILKS, Dr, case of fractured spine and herpes zoster	. 128
—— on abeyance of the will in hysteria	. 207, 217, 229
—— on nervous mimicry	. 212
—— on peripheral facial palsy	. 253
—— on stunning of the cord	33, 148
—— on the correct estimation of pain	. 248
—— on the value of spinal pain as a symptom	. 142
Will, abeyance of, in hysteria	. 207, 217, 229
—— the, in hypnotism	. 219
—— weakness of the, Paget upon	. 230
Witness, the medical	. 292
WRENCH, Mr, on the effects of fixed idea, a case	. 194

X

Xanthelasma palpebrarum	. 226

Y

Young, influence of shock on the	. 169

Z

Zone, of hyperæsthesia	78, 14
Zoster, herpes	. 128

INDEX OF AUTHORS

Abercrombie, 9, 34, 47
Althaus, 225
Bastian, 52
Bell, 10, 42, 68
Boyer, 12
Briquet, 215
Brodie, 3, 5, 7
Buzzard, 128, 132, 137, 224, 289
Charcot, 227
Clark, Le Gros, 109, 110
Clarke, Bruce-, 210
Clarke, Lockhart, 101
Clinton, 294
Cooper, 70
Curling, 46
Donders, 180
Edmunds, 22
Erb, 111, 207
Erichsen, 58, et seq., 217
Ferrier, 26
Fletcher, 337
Fox, 96
Gavin, 250
Gowers, 137, 141, 153, 206, 209
Gull, 40, et seq.
Hammond, 206
Heidenhain, 222, 242
Hewett, 24
Hilton, 126, 248
Hodges, 156
Hood, 121
Hunter, 47

Hutchinson, 24, 27, 38, 48, 181, 186, 226, 251
Johnson, 115
Jones, Handfield, 178, 218
Jordan, 160, 163, 169, 185, 188
Kemble, 197
Laycock, 119, 220
Lidell, 15, 16, 35
Liveing, 205
Lochner, 46
Lush, 107
Mansell-Moullin, 160
Maudsley, 192, 194, 197, 201, 214, 277
Mayo, 10, 43
Moxon, 228
Ogston, 247
Paget, 195, 204, 214, 225, 230
Petit, 151
Putnam, 274
Rigler, 113
Reynolds, 267, 293
Ross, 138
Savory, 36
Sharkey, 50
Shaw, 39, 59, 103, 106, 117, 143
Spence, 29
Sully, 192, 194
Syme, 14
Tuke, 164, 196, 207
Wilks, 33, 128, 142, 207, 212, 217, 229, 248, 253
Wrench, 194

PRINTED BY
J. E. ADLARD, BARTHOLOMEW CLOSE.

www.ingramcontent.com/pod-product-compliance
Lightning Source LLC
Chambersburg PA
CBHW022114290426
44112CB00008B/672